# *Strange Attractors*

# Strange Attractors

## Poems of Love and Mathematics


Edited by
Sarah Glaz
JoAnne Growney



CRC Press
Taylor & Francis Group
Boca Raton London New York

CRC Press is an imprint of the
Taylor & Francis Group, an **informa** business
AN AK PETERS BOOK

Editorial, Sales, and Customer Service Office

A K Peters, Ltd.
888 Worcester Street, Suite 230
Wellesley, MA 02482
www.akpeters.com



Library of Congress Cataloging-in-Publication Data

Strange attractors : poems of love and mathematics / edited by Sarah Glaz, JoAnne Growney.
p. cm.
Includes bibliographical references and index.
ISBN 978-1-56881-341-7 (alk. paper)
1. Mathematics--Poetry. 2. Love--Poetry. 3. American poetry--21st century. I. Glaz, Sarah, 1947- II. Growney, JoAnne Simpson.
PS595.M36S73 2008
808.81'93543--dc22

2008017984

Cover image: Binary Heart by Randall Munroe (http://xkcd.com/99/). Used by permission.

Reprinted 2010 by CRC Press
CRC Press
6000 Broken Sound Parkway, NW
Suite 300, Boca Raton, FL 33487
270 Madison Avenue
New York, NY 10016
2 Park Square, Milton Park
Abingdon, Oxon OX14 4RN, UK

# Contents

Introduction xi
Mathematical Poetry Resources xvii

1 Romantic Love: *from Heartaches to Celebrations*

Diane Ackerman, from *Tales from A Sonnetarium: III, Kismet* 3
Manuel Alegre, from *Polar Heart* 4
Yehuda Amichai, from *Israeli Travel: Otherness is All, Otherness is Love* 5
Amy Quan Barry, *If* $\frac{dy}{dx} = \frac{4x^3 + x^2 - 12}{\sqrt{2x^2 - 9}}$, *then* 6
Judith Baumel, *Fibonacci* 7
Lucian Blaga, *Tally* 9
David Brooks, *Yes* 10
Elizabeth Barrett Browning, *How do I love thee? Let me count the ways.* 11
Gaius Valerius Catullus, *Let's Live and Love: To Lesbia* 12
Siv Cedering, *The Acrobat* 13
Inger Christensen, from *alphabet: love* 14
J. V. Cunningham, *Statistics* 15
Jon Davis, *The Accounting* 16
Jean de Sponde, *Sonnet of Love XIII* 17
Emily Dickinson, *It's all I have to bring today—* 18
Emily Dickinson, *We shall find the Cube of the Rainbow.* 19
John Donne, *Love's Growth* 20
John Donne, *The Computation* 21
Gerald Early, *With Linnet and Rosalind on the Evening of the Equinox (or an Exposition on the Big Bang Theory)* 22
Jennifer Granville, *Cat in a Box* 23
Emily Grosholz, *The Ratio of Green* 24
JoAnne Growney, *Floating* 25

Miriam Hammel, *Entropy* 26
H. L. Hix, from *Orders of Magnitude* 27
Tony Hoagland, *Self-Improvement* 29
Roald Hoffman, *Why Does Disorder Increase in the Same Direction of Time as That in Which the Universe Expands?* 31
Jonathan Holden, *Sex and Mathematics* 32
X. J. Kennedy, *Geometry* 34
Jacqueline Lapidus, *Several Hypotheses and a Proposition* 35
Philip Larkin, *Counting* 37
Stanislaw Lem, from *The Cyberiad* 38
Marjorie Maddox, *Venn Diagrams* 40
Andrew Marvell, *The Definition of Love* 42
Harry Mathews, *Safety in Numbers* 44
Harry Mathews, *Shakespeare on Love* 46
Peter Meinke, *3.141592 . . .* 47
Peter Meinke, *Distances* 48
Dunya Mikhail, *The Theory of Absence* 49
Deanna Nikaido, *July 18, 2005* 50
Bin Ramke, from *Gravity & Levity* 51
Pattiann Rogers, *On Your Imminent Departure: Considering the Relative Importance of Various Motions* 53
Becky Dennison Sakellariou, *Math Is Beautiful and So Are You* 54
Eryk Salvaggio, from *Five Poems about Zero* 55
Maureen Seaton, *The Freezing Point of the Universe* 56
King Solomon, from *The Song of Songs* 57
C. K. Stead, from *Walking Westward* 58
Amy Uyematsu, *The Meaning of Zero: a Love Poem* 60
Amy Uyematsu, *Unfinished Calculations* 61
Andrei Voznesensky, *Parabolic Ballad* 63
William Wall, *Mathematics: Insects* 65
Ronald Wallace, *Chaos Theory* 66
John Wheelwright, *Two Tongues in a Tower* 68
Yang Mu, *Geometry: River Goddess* 69

## 2 Encircling Love: *of Family, Nature, Life, and Spirit*


Rafael Alberti, *Harold Lloyd, Student* 73
Karren LaLonde Alenier, *Dialectic of the Census Takers* 75
Dante Alighieri, from *The Paradiso: Canto XXXIII* 76
Carl Andre, *On the Sadness* 77
Linda Bierds, *Counting: Gregor Mendel in the Prelacy* 79
Robin Chapman, *Nonlinear Function* 80
Edward E. Chipman, *God Loves a Curve* 81
Dan Clark, *Practical Application* 82
Jennifer Crow, *Mathematics* 83
Kathryn DeZur, *Fibonacci Numbers* 85
Rita Dove, *Flash Cards* 86
Cathryn Essinger, *My Dog Practices Geometry* 87
Robert Frost, *Moon Compasses* 89
James Galvin, *Geometry Is the Mind of God* 90
Sarah Glaz, *Love Story* 91
Emily Grosholz, *Proportions of the Heart* 92
Emily Grosholz, *The Shape of Desire* 94
Bob Grumman, *Mathemaku No. 10* 95
Israel Har, *The Zero* 96
Jonathan Holden, *The Departure of an Alphabet* 97
Langston Hughes, *Addition [1]* 99
Solomon Ibn Gabirol, from *The Kingly Crown: The Praises of God* 100
James Kirkup, *How to Find the Longest Distance Between Two Points* 101
Stanley Kunitz, *Geometry of Moods* 102
John Stephen Lew, *The Comet* 103
Israel Lewis, *I Find My Faith in the Flatness of Space* 104
Henry Lok, from *Sundry Christian Passions: Square Poem in Honor of Elizabeth I* 105
Edwin Markham, *Outwitted* 107
Wilmer Mills, *An Equation for My Children* 108
Agi Mishol, *Geese* 109
Wendy Mnookin, *Math* 110

Kyoko Mori, *Barbie Says Math Is Hard* 111
Carl Sandburg, *Child Margaret* 112
Shel Silverstein, *Smart* 113
Young Smith, *She Considers the Dimensions of Her Soul* 114
Elizabeth Anne Socolow, *Asymptotes* 115
A. E. Stallings, *Sine Qua Non* 116
Mark Strand, *The New Poetry Handbook* 117
Stephanie Strickland, *True North 1* 119
Stephanie Strickland, *Who Counts, Counts* 120
John Updike, from *Midpoint: The Play of Memory* 122
Ramón López Velarde, *Loyal, My Heart . . .* 123
John Vieira, *The Lake Swan, the Tom* 124

3 *Unbounded Love: for Mathematics and Mathematicians*

Rafael Alberti, *To Divine Proportion* 127
Sandra Alcosser, *My Number* 128
Jakob Bernoulli, from *Treatise on Infinite Series* 130
Bhaskaracharya, from *Lilavati* 131
Ray Bobo, *Give Me an Epsilon and I Will Treat It Well* 132
Ann Calandro, *Where the Kissing Never Stops* 133
Lewis Carroll, from *The Hunting of the Snark: The Beaver's Lesson* 134
Hanns Cibulka, *Mathematics* 136
Marion Deutsche Cohen, *Scared and the Intermediate Value Theorem* 137
Marion Deutsche Cohen, *What Drove Me into Math* 138
Mary Cornish, *Numbers* 139
Jonathan Coulton, *Mandelbrot Set* 141
Keith Allen Daniels, *Mathrimony* 143
E.P. Dempster, from *A Mathematical Madness* 145
Theodore Deppe, *The Proof* 146
James Dickey, *Math* 148
Rita Dove, *Geometry* 149
John L. Drost, *I Even Know of a Mathematician* 150

Cathryn Essinger, *The Mathematician, Counting* 151
Cathryn Essinger, *You Are Right* 152
Sarah Glaz, *Calculus* 153
Sarah Glaz, *Late Afternoon at the Workshop on Commutative Rings* 155
Douglas Goetsch, *Counting* 156
JoAnne Growney, *My Dance Is Mathematics* 158
JoAnne Growney, *San Antonio, January, 1993* 160
Haipeng Guo, *When a P-Man Loves an NP-Woman* 161
Robert Hass, *"... White of Forgetfulness, White of Safety"* 163
J. A. H. Hunter, *A Lady and Her Lingerie* 165
Bob Kurosaka, *There Was a Young Maiden* 166
Tom Lehrer, *There's a Delta for Every Epsilon* 167
Kaz Maslanka, *Sacrifice and Bliss* 168
James Clerk Maxwell, *A Lecture on Thomson's Galvanometer* 169
John Ward McClellan, *A Lady of 80* 170
Randall Munroe, *Useless* 171
Howard Nemerov, *Figures of Thought* 172
Pablo Neruda, *Ode to Numbers* 173
Katharine O'Brien, *Undefined Terms* 176
Katharine O'Brien, *Valentine* 177
Bill Parry, *Alexander's Horned Sphere* 178
Bill Parry, *Argument* 179
Raymond Queneau, *Swigns* 180
William J. Macquorn Rankine, *The Mathematician in Love* 181
Len Roberts, *We Sat, So Patient* 183
Carl Sandburg, *Number Man* 185
Ed Seykota, from *Borderline—A Fractal Poem* 186
Frederick Soddy, *The Kiss Precise* 188
Thorold Gosset, *The Kiss Precise (Generalized)* 189
Helen Spalding, *Let Us Now Praise Prime Numbers* 190
Stephanie Strickland, *Striving All My Life* 191
Michael Stueben, *Valentine* 192
Unknown author, *Puzzle* 193

Alissa Valles, *Mathematician* 194
Ellen Wehle, *The Song of 10* 196
William Carlos Williams, *Aigeltinger* 197
Clarence R. Wylie, Jr., *Paradox* 198

Contributors' Notes 199
About the Mathematicians 221
Acknowledgments 229
Index 243

# Introduction

We, the editors who have gathered these poems, are mathematicians and poets. This collection has been an opportunity for us to merge our two interests into one project and share it with you, the reader. These pages invite you to explore the ways that mathematical ideas inhabit poetry.

The poems collected here represent the best poetry has to offer—with work by the world's finest poets—including Nobel and Pulitzer Prize winners and Poet Laureates. This collection is international, including translations of major poetic voices. The time spread for the poems is at least 3000 years—we include a fragment of *The Song of Songs* by King Solomon, circa 1000 BC, as well as contemporary American poetry. The common theme is *love*, and we interpret this theme with broad universality, from the peaks and valleys of romantic love, through the encircling love of family, nature, life and spirit, to the love that focuses on mathematics and mathematicians. Each poem has a strong link to mathematics in content, form, or imagery, and the collection engages a variety of mathematical topics—from counting to commutative rings, from the intermediate value theorem to infinity. A number of eminent mathematicians appear in the poems and highlight the human history of mathematics; a collection of biographical notes for these figures follows the poetry.

The Mathematical Poetry Resources section following this Introduction includes references for selected publications and other material connecting poetry and mathematics, as well as links to online resources for those wishing to explore various mathematical terms or topics.

Since ancient times poets have used mathematical imagery and ideas in their verses, and mathematicians versified when the spirit moved them. The earliest poet who is known by name is also the earliest mathematician who is known by name: Enheduanna, chief priestess of the moon god Nanna, in the city of Ur, daughter of the Akkadian king Sargon who captured and ruled over the region of Sumer during 2334–2279 BC. Enheduanna's duties as chief priestess required mathematics—she was involved in astronomical calendrical calculations ("measuring the skies") and also in civic mathematics related to setting boundaries of properties and engineering measurements for construc-

tion ("placing the measuring cord"). Her duties as religious leader also included the composition of temple hymns to a variety of deities. Here are lines from one of her hymns dedicated to Nisaba's temple in Ereš. Nisaba, the grain goddess, was patron of scribal arts and mathematical calculations:

> The true woman who possesses exceeding wisdom,
> She consults a tablet of lapis lazuli,
> She gives advice to all lands,
> She measures off the heavens, she places the
> measuring-cords on the earth.
> Nisaba, praise![1]

Well before the start of the twentieth century, the amount of available knowledge exceeded the learning capacity of any single individual. The result was specialization, and with it separation, sometimes perceived as rift, between mathematics and the sciences on one side and the arts and humanities on the other. In spite of this, there has been a constant stream of people on both sides of the divide working to bridge the gap.

As we look back toward and into the twentieth century, we see several forebears to our volume—people and publications that, despite the explosion of information and knowledge, have kept alive and strengthened the connection between mathematics and poetry.

Gathered by Robert E. Moritz, *Memorabilia Mathematica* (1914) is a collection of 1140 anecdotes, aphorisms, and passages by noted mathematicians, scientists, and writers on a broad variety of mathematical topics—small fragments from a selection of poetry by writers such as Dante, Goethe, and Tennyson are included, as well as poetic lines from several mathematicians. Moritz's focus, however, is on the mathematical content of verses and not on the poems.

Another landmark is Scott Buchanan's extended essay *Poetry and Mathematics*, which first appeared in print in 1929. In this philosophical treatise, the author traces with dexterity parallel patterns that he has observed in these two disciplines. *Poetry and Mathematics* is dedicated to Dante and Kepler—*to the poet who was a mathematician and to the mathematician who was a poet.*

Two fine and very different anthologies appeared in the 1950s, *Imagination's Other Place: Poems of Science and Mathematics*, edited

[1]Translated from Sumerian cuneiform sources by Åke W. Sjöberg and E. Bergmann S.J.

by Helen Plotz, and *Fantasia Mathematica,* edited by Clifton Fadiman. Plotz's anthology contains only poems and emphasizes work from well-known literary figures, whereas Fadiman's mathematical literature collection is largely prose—with its verse tending toward the light and humorous varieties. Fadiman's anthology became popular and was followed in 1962 by an equally popular sequel: *The Mathematical Magpie,* both reprinted in 1997. Following the pattern of Fadiman's anthologies, William Frucht edited *Imaginary Numbers,* published in 1999. In the vein of Plotz's anthology of science poems are the two collections *Songs from Unsung Worlds: Science in Poetry,* edited by Bonnie Bilyeu Gordon (1985), and *Verse & Universe: Poems about Science and Mathematics,* edited by Kurt Brown (1998). Our collection, consisting entirely of mathematical poems, mimics the broad variety of these volumes—with lyrics from silly to serious, from bawdy to beatific—and our emphasis has tended toward work from strong poetic voices.

*Against Infinity: An Anthology of Contemporary Mathematical Poetry,* edited by Ernest Robson and Jet Wimp (1979), is, to our knowledge, the earliest anthology to include only mathematical poetry. This slim volume, a treasure for mathematical poetry lovers, introduces mathematicians to concrete poetry and to poems constructed using mathematical symbols (sometimes *only* mathematical symbols). Our anthology, about thirty years later, is several times larger than *Against Infinity* and is the first-ever anthology of mathematical *love* poems.

Approaches to mathematics that reach across cultural divides to history, literature and the arts have sometimes been termed "humanistic mathematics." A mathematician who was a leader in the spread of humanistic mathematics is Alvin M. White, Professor Emeritus at Harvey Mudd College. White established and edited the *Humanistic Mathematics Network Newsletter/Journal* from 1987 to 2004. JoAnne served as poetry editor on this journal and also edited *Numbers and Faces,* a small anthology of mathematical poetry published by the HMNJ in 2001 that featured work by an international collection of poets. In 2002, Sarah's first mathematical poem, "Calculus" (included in this volume), was published in Volume 26 of HMNJ.

During the thirty-year period 1956–1986, Martin Gardner's "Mathematical Games" column for *Scientific American* frequently brought poetry to the attention of mathematicians. Gardner became well-known for his ability to make arcane topics accessible and enjoyable—and he was one of the popularizers of the *OULIPO* literary movement, formed in the 1960s in France by a group of writers

and mathematicians who developed ingenious strategies (often mathematical) for generation of new literature. The work of two Oulipians (Harry Mathews and Raymond Queneau) is included in our collection. Gardner, who occasionally versifies, has supported our anthology from the start and helped us locate the copyright-holder of one of our poems. He has offered no mathematical "love poem" of his own for us to include, but we do have from him this quatrain:

> $\pi$ goes on and on,
> And e is just as cursed.
> I wonder, how does $\pi$ begin
> When its digits are reversed?

Poetry journals as well as mathematical ones do, from time to time, include mathematical poems. For example, six of the poems of this anthology first appeared in the national journal *POETRY*. Additionally, a number of poets in this volume have had work presented in one of The Mathematical Association of America's journals: *The American Mathematical Monthly, The Mathematics Magazine,* and *The College Mathematics Journal,* as well as in *The Mathematical Intelligencer* (Springer).

There are additional new developments in humanistic mathematics: in 1998, the annual international conference of Bridges: Mathematical Connections in Art, Music, and Science was created and has met each year since inception. During the conference eclectic groups of mathematicians, scientists, artists, educators, musicians, writers, computer scientists, sculptors, dancers, weavers, and model builders come together in an atmosphere of mutual exchange and encouragement. In 2006, interested members of the Mathematical Association of America formally organized as SIGMAA–ARTS, a special interest group that fosters exchange and connections between mathematics and the arts. Established in 2007, the *Journal of Mathematics and the Arts* (Taylor & Francis) is a peer review journal that focuses on links and commonalities between mathematics and the arts.

As teachers who have used the literature of mathematics in our classrooms, we hope that this collection will be useful to other mathematicians, poets, and teachers to enliven or complement their courses. We intend this collection to be a special gift—and hope that these poems will offer unbounded opportunity for readers' pleasure.

We are indebted to all of the poets whose poems are collected here for creating fine verse. We thank both poets and publishers and other permission grantors for generosity of spirit as we negotiated for the

rights to reprint poems. We are grateful to the University of Connecticut's Office of Sponsored Programs, the College of Liberal Arts and Sciences, and the Department of Mathematics, and also to A K Peters, Ltd., for grants that enabled us to obtain rights to reprint poems in this volume. We wish to thank Klaus Peters, our publisher, and Charlotte Henderson, our editor, at A K Peters, Ltd., who believed in our project from the start, and who continually encouraged and assisted us into bringing it to completion. Most of all, we express our appreciation to each other—for kindness, forbearance, and forgiveness as we shared the burdens and the joys of putting this volume together.

In addition—

Sarah Glaz would like to thank her family, friends, and colleagues who encouraged her in this undertaking: her mother, Amalia Dauer, for reading poetry to her before she learned to count; her husband, Joe Glaz, and her writer-mathematician son, Ron Glaz, for being there and cheering her on through the ups and the downs; Marc Rubenstein, Claudine Smith, Riva and Arnold Berleant, and Vera Schwarcz, for years of friendship, conversation and poetry; her colleagues, Miki Neumann, Joan and Stu Sidney, and Mel Hochster for poems and support. She also is grateful for memories of her father, Philip Dauer, who supported her interest in mathematics.

JoAnne Growney would like to thank her family and friends for loving support throughout this project. Her warm gratitude is extended also to members of a growing network—in Germany and Portugal, in New Zealand and India and Romania, in Bloomsburg and Silver Spring—of friends who love both poetry and mathematics and who exchange mathematical poems with each other. Special thanks also to Marjorie Maddox and Jerry Wemple—experienced anthologists who offered valuable suggestions. JoAnne's portion of this anthology is dedicated to her grandchildren—Carly, Shayla, Mika, Emma, Daniel, Serena, and Ami—whose current mathematical verses are "counting rhymes," but who one day will discover the poetry of theorems and the power of poems . . . .

*Sarah Glaz*
*Storrs, Connecticut*

*JoAnne Growney*
*Silver Spring, Maryland*

*February 2008*

# Mathematical Poetry Resources for Further Exploration

## Explanations of Mathematical Concepts

*Wolfram Math World*, Eric Weisstein, Wolfram Research: www.mathworld.wolfram.com.

*The Mathematical Atlas*, Dave Rusin, Northern Illinois University: www.math-atlas.org.

## Anthologies of Poetry of Mathematics and the Sciences

*Against Infinity: An Anthology of Contemporary Mathematical Poetry*, eds. Ernest Robson and Jet Wimp, Primary Press, Parker Ford, PA, 1979.

*Imagination's Other Place: Poems of Science and Mathematics*, ed. Helen Plotz, Thomas Y. Crowell Company, New York, 1955.

*Numbers and Faces: A Collection of Poems with Mathematical Imagery*, ed. JoAnne Growney, Humanistic Mathematics Network, Claremont, CA, 2001.

*Songs from Unsung Worlds: Science in Poetry*, ed. Bonnie Bilyeu Gordon, Birkhäuser, Boston, 1985.

*Verse & Universe: Poems about Science and Mathematics*, ed. Kurt Brown, Milkweed Editions, Minneapolis, 1998.

## Anthologies of Mathematical Literature That Include Poems

*Memorabilia Mathematica*, ed. Robert E. Moritz, Macmillan, New York, 1914. Reprinted by The Mathematical Association of America, Washington, DC, 1993.

*Fantasia Mathematica*, ed. Clifton Fadiman, Simon and Schuster, New York, 1958. Reprinted by Copernicus, an imprint of Springer, New York, 1997.

*The Mathematical Magpie*, ed. Clifton Fadiman, Simon and Schuster, New York, 1962. Reprinted by Copernicus, an imprint of Springer, New York, 1997.

*Imaginary Numbers*, ed. William Frucht, John Wiley & Sons, New York, 1999.

## About Mathematical Poetry: Selected Articles and Collections

*Discovering Patterns in Mathematics and Poetry*, Marcia Birken and Anne C. Coon, Rodopi, Amsterdam-New York, 2008.

*Humanistic Mathematics Network Newsletter /Journal*, 1987–2004. Established and edited by Alvin White, Harvey Mudd College; online editors Sandra and Philip Keith, St. Cloud State College (2002–2004). Selected issues archived: http://www2.hmc.edu/www_common/hmnj.

*Martin Gardner's Mathematical Games: The Entire Collection of His Scientific American Column*, on CD, The Mathematical Association of America, Washington, D.C., 2005.

"Mathematics in Poetry," JoAnne Growney, *Journal of Online Mathematics and its Applications*, Article ID: 1262, October 2006.

"Mathematics Influences Poetry," JoAnne Growney, *Journal of Mathematics and the Arts*, Vol. 2, No.1, pp. 1–7, 2008.

*OULIPO: A Primer of Potential Literature*, Translator and Editor, Warren F. Motte, Jr., Dalkey Archive Press, Urbana-Champaign, 1998.

*Poetry and Mathematics*, Scott M. Buchanan, The John Day Company, New York, 1929. Reprinted by The University of Chicago Press, Chicago, 1975.

"Polyaesthetics and Mathematical Poetry," Kaz Maslanka, *Journal of Mathematics and the Arts*, Vol. 1, pp. 35–40, 2007.

*The World of Mathematics: A Small Library of the Literature of Mathematics from A'h-mosé the Scribe to Albert Einstein, Presented with Commentaries and Notes* by James R. Newman, 4 volumes, Simon and Schuster, New York, 1956. Reprinted by Dover Publications, New York, 2003.

"Versed in Math: Modeling with Poetry in an Intermediate Algebra Course and Beyond," Sarah Glaz and Su Liang, preprint.

## Organizations Supporting Mathematics in the Arts

The Bridges Organization: Mathematical Connections in Art, Music, and Science: www.bridgesmathart.org.

SIGMAA–ARTS: Special Interest Group on Mathematics and the Arts, The Mathematical Association of America, Washington, D.C.: www.maa.org/sigmaa/arts.

## *Poems in This Collection*

The Acknowledgments section at the end of this volume contains information about mathematical poetry books published by poets featured in this volume. In addition, mathematical poetry is available on the Internet, including sites maintained by many mathematical poets and songwriters. Search engines will locate up-to-date URLs for each.

## *Editor–Publisher Websites*

Sarah Glaz: www.math.uconn.edu/~glaz.

JoAnne Growney: http://joannegrowney.com.

A K Peters, Ltd.: www.akpeters.com.

Part 1

# Romantic Love *from* Heartaches to Celebrations

# Diane Ackerman

## From *Tales from A Sonnetarium: III, Kismet*

*What can't be said can't be said,*
*and can't be whistled either.*
*—Wittgenstein*

Wittgenstein was wrong: when lovers kiss
they whistle into each other's mouth
a truth old and sayable as the sun,
for flesh is palace, aurora borealis,
and the world is all subtraction in the end.

The world is all subtraction in the end,
yet, in a small vaulted room at the azimuth
of desire, even our awkward numbers sum.
Love's syllogism only love can test.

But who would quarrel with its sprawling proof?
The daftest logic brings such sweet unrest.
Love speaks in tongues, its natural idiom.
Tingling, your lips drift down the xylophone
of my ribs, and I close my eyes and chime.

## Manuel Alegre

### From *Polar Heart*

I do not know what colour the ships are
when they shipwreck in your arms
I know that there is a body never found out there in the sea
and that living body is your immaterial body
your promise in the masts of all sailboats
the perfumed isle of your legs
your womb of conches and coral
the grotto where you await me
with your lips of foam and saltiness
your shipwrecks
and the great equation of the wind and the voyage
where chance flowers with its mirrors
its signs of rose and discovery.

I do not know the colour of that line
where the moon and masts cross
but I know that in each road there is a corner
an opening between the routine and the marvelous
there is an hour of fire for blue
the hour in which I find and do not find you
there is an opposite angle
a magical geometry where everything might be possible
there is an imaginary sea open on each page
don't you tell me that routes will
never again be born of desire
and I want the southern cross of your hands
I want your name written in the tides
in this city where in the most absurd place
at a no entry sign or traffic lights
all sunsets tell me who you are.

Translated from the Portuguese by Mick Greer and Graça Margarido

## From *Israeli Travel: Otherness is All, Otherness is Love*

I remember a problem in a math book
about a train that leaves from place A and another train
that leaves from place B. When will they meet?
No one ever asked what happens when they meet:
Will they stop, or pass each other, or collide?
None of the problems was about a man who leaves from place A
and a woman who leaves from place B. When will they meet,
will they meet at all, and for how long?
As for that math book: Now I've reached
the final pages with the answers.
Back then it was forbidden to look.
Now it is permitted. Now I check
where I was right and where I went wrong,
and know what I did well and what I did not do. Amen.

Translated from the Hebrew by Chana Bloch and Chana Kronfeld

Amy Quan Barry

## If $\frac{dy}{dx} = \frac{4x^3 + x^2 - 12}{\sqrt{2x^2 - 9}}$, *then*

you are standing at the ocean,
in the moon's empirical light
each mercurial wave

like a parabola shifting on its axis,
the sea's dunes differentiated & graphed.
*If this, then that.* The poet

laughs. She wants to lie
in her own equation, the point slope
like a woman whispering *stay me*

*with flagons.* What is it to know the absolute value
of negative grace, to calculate
how the heart becomes the empty set

unintersectable, the first & the last?
But enough.
You are standing on the shore,

the parameters like wooden stakes.
Let $x$ be the moon like a notary.
Let $y$ be all things left unsaid.

Let the constant be the gold earth
waiting to envelop what remains,
the sieves of the lungs like two cones.

## *Fibonacci*

*for Abraham Baumel*

Call it windfall

finding your calculation

come, finally,
to the last decimal point of pi.

In the silence of January snow
a ladybug survives the frost
and appears on the windowpane.

She crawls a tiny space.
Hesitant. Reverses. Forward,
like a random-number generator,
the walking computer frog
who entertains mathematicians.

Think of the complexity
of temperature, quantification
of that elusive quality "heat."
Tonight, for instance,
your hands are colder than mine.
Someone could measure
more precisely than we
the nature of this relationship.

Learn the particular strength
of the Fibonacci series,
a balanced spiraling
outward of shapes,
those golden numbers
which describe dimensions
of sea shells, rams' horns,
collections of petals

and generations of bees.
A formula to build
your house on,
the proportion most pleasing
to the human eye.

# *Tally*

I tally in the ancient way.
I count like the shepherd
how many white, how many black
—days, all year round.

I count the steps, of the beautiful one,
to the threshold of the door.
I count how many stars there are
in the nest of the Mother Hen.

However many, the lot—I count,
smoke and illusions,
the whole day,—count, count
roads and missed ways.

I count the stones on which
she crosses the ford, that beauty—
and all the sins for which
hell will surely burn me.

Translated from the Romanian by Brenda Walker and Stelian Apostolescu

## David Brooks

# Yes

You know how
just before we die
our whole life is supposed
to flash before our eyes?
Well, should that happen, we'd surely have,
while it is flashing,
to come to that moment when our whole life
flashes before our eyes,
and while that was happening
all over again, we'd come again
to that same moment, and so on,
which is only to say that, while I know
this might not have been good enough for Zeno
and that it's a certainty
that death happens anyway, for that one
minute, when that thought came
and I imagined
living this life
over and over,
I said to myself, despite
all the effort, all the
pain of it,
despite
all that has happened
and is likely to
again and again,
Yes, I thought, as I was watching you
getting ready for bed tonight,
Yes, though I knew
even then
it was crazy beyond measure,
*Yes*, I would, *Yes*,
*Please*, *Yes*.

## How do I love thee? Let me count the ways.

How do I love thee? Let me count the ways.
I love thee to the depth and breadth and height
My soul can reach, when feeling out of sight
For the ends of Being and ideal Grace.
I love thee to the level of everyday's
Most quiet need, by sun and candle-light.
I love thee freely, as men strive for Right;
I love thee purely, as they turn from Praise.
I love thee with the passion put to use
In my old griefs, and with my childhood's faith.
I love thee with a love I seemed to lose
With my lost saints,—I love thee with the breath,
Smiles, tears, of all my life!—and, if God choose,
I shall but love thee better after death.

## Let's Live and Love: To Lesbia

Let us live, my Lesbia, let us love,
and all the words of the old, and so moral,
may they be worth less than nothing to us!
Suns may set, and suns may rise again:
but when our brief light has set,
night is one long everlasting sleep.
Give me a thousand kisses, a hundred more,
another thousand, and another hundred,
and, when we've counted up the many thousands,
confuse them so as not to know them all,
so that no enemy may cast an evil eye,
by knowing that there were so many kisses.

Translated from the Latin by A. S. Kline

# *The Acrobat*

There is no bravery in flight.
    There are only laws
        Of four-dimensional geometry
And a body tracing spiral and arc
    Of a predetermined trajectory
        That sends me off with ease
Leaving the platform to leap
    From trapeze to trapeze.

It is not the gravity of earth
    Pulling all things closer
        That scares me,
But the gravity of love that attracts
    And urges me to let go,
        To trust that you will catch me,
The way practice has taught me to trust
    The sky, my brother, and combined mastery.

So if I dared to stand
    Barefoot on the ground before you,
        My face close to your face,
Hand to hand, a fear of heights
    Would somersault inside me
        At the thought of leaping out
Into the space your eyes offer
    To hold me.

# Inger Christensen

## From *alphabet: love*

love exists, love exists
your hand a baby bird so obliviously tucked
into mine, and death impossible to remember,
impossible to remember how inalienable
life, as easily as chemicals drifting
over the knotgrass and rock doves, all of it
is lost, vanishing, impossible to remember that
there and there flocks of rootless

people, livestock, dogs exist, are vanishing;
tomatoes, olives vanishing, the brownish
women who harvest them, withering, vanishing,
while the ground is dusty with sickness, a powder
of berries and leaves, and the buds of the caper
are never gathered, pickled with salt
and eaten; but before they vanish, before we
vanish, one evening we sit at the table with
a little bread, a few fish without cankers, and water
cleverly turned into water, one of
history's thousands of war paths suddenly
crosses the living room, you get up, limits,
given limits exist, streets, oblivion

everywhere, but ...

Translated from the Danish by Susanna Nied

Note from the translator: The length of each section of Inger Christensen's *alphabet* is based on Fibonacci's sequence, a mathematical sequence beginning 0, 1, 1, 2, 3, 5, 8, 13, 21..., in which each number is the sum of the two previous numbers.

## *Statistics*

They both collected for the March of Dimes,
Just middle-aged and average. As they should,
They had coitus two-and-six-tenths times
A week. The six-tenth time was not so good.

Jon Davis

## *The Accounting*

Numbers scraping their heels all night in the attic above them.
The accounting was necessary but brought them near death.
Like a fountain emptying itself for the tourists.
Couldn't the woman in the restaurant in the booth in the semi-private
room open her shirt?
In one view the numbers represented his errors of judgment.
In another, less sanguine view, his urge not to die into the dailiness.
The dailiness with its postponements, its sacks and surcharges, its
relentless taxation.
Couldn't the woman lean against the bed's metal rails?
The punishment was merely a distraction from the real punishment
which was distraction.
Couldn't the woman suddenly before the meal was complete?
Couldn't she kiss him press him against the kitchen counter?

# *Sonnet of Love* XIII

"Give me a place to stand," Archimedes said,
"and I can move the world." Paradoxical, clever,
his remark which first explained the use of the lever
was an academic joke. But if that dead

sage could return to life, he would find a clear
demonstration of his idea, which is not
pure theory after all. That putative spot
exists in the love I feel for you, my dear.

What could be more immovable or stronger?
What becomes more and more secure, the longer
it is battered by inconstancy and the stress

we find in our lives? Here is that fine fixed point
from which to move a world that is out of joint,
as he could have done, had he known a love like this.

Translated from the French by David R. Slavitt

## *It's all I have to bring today–*

It's all I have to bring today—
This, and my heart beside—
This, and my heart, and all the fields—
And all the meadows wide—
Be sure to count—should I forget
Some one the sum could tell—
This, and my heart, and all the Bees
Which in the Clover dwell.

## *We shall find the Cube of the Rainbow.*

We shall find the Cube of the Rainbow.
Of that, there is no doubt.
But the Arc of a Lover's conjecture
Eludes the finding out.

# *Love's Growth*

I scarce believe my love to be so pure
As I had thought it was,
Because it doth endure
Vicissitude, and season, as the grass;
Methinks I lied all winter, when I swore,
My love was infinite, if spring make it more.
But if this medicine, love, which cures all sorrow
With more, not only be no quintessence,
But mixed of all stuffs, paining soul, or sense,
And of the sun his working vigour borrow,
Love's not so pure, and abstract, as they use
To say, which have no mistress but their Muse,
But as all else, being elemented too,
Love sometimes would contemplate, sometimes do.

And yet not greater, but more eminent,
Love by the Spring is grown;
As in the firmament,
Stars by the sun are not enlarged, but shown,
Gentle love deeds, as blossoms on a bough,
From love's awakened root do bud out now.
If, as in water stirred more circles be
Produced by one, love such additions take,
Those like so many spheres, but one heaven make,
For, they are all concentric unto thee,
And though each spring do add to love new heat,
As princes do in times of action get
New taxes, and remit them not in peace,
No winter shall abate the spring's increase.

## The Computation

For the first twenty years, since yesterday,
  I scarce believed thou couldst be gone away,
For forty more, I fed on favours past,
  And forty on hopes, that thou wouldst, they might last.
Tears drowned one hundred, and sighs blew out two,
  A thousand, I did neither think, nor do,
  Or not divide, all being one thought of you;
  Or in a thousand more, forgot that too.
Yet call not this long life; but think that I
Am, by being dead, immortal; can ghosts die?

## Gerald Early

# *With Linnet and Rosalind on the Evening of the Equinox (or an Exposition on the Big Bang Theory)*

"Stand," she said, in the night air, "stand still as a star,"
As if a star were ever still or standing, but a throbbing
Of light, a zoomed memory of its source when we see it there;
And there she looks, in the night sky, pockmarked, projected,
Punctured, the mathematics of a gun universe, looks up, at an angle,
Imagining, I suppose, the starlight something like a balance beam
That, hot and narrow, she could walk on to heaven as a heroine,
Convulsive as lightning, straight as an equation, guileless as girlhood.

What to make of the warmth of this evening, through one's shoes,
A germinating rush, the hatching, the release of life, like gunshots,
A cracked-open world, wasted multiplicity, as the other she, looking
down,
Imagining the eruptive hermitages of coitus, o captious seer,
Says, "Vibrate, divide, work like an insect," and so, pointilistically
Trilling, the crickets announce, to the man startled by that richness
In his girls so much matching there another fertility, the
Return of the reign of sexed-up nature, the rites of another
Passage that, in passing, heeds not the unheated, this unbidden
Uncoiling carnival and carnage of regeneration.

Jennifer Granville

## Cat in a Box

*A poem dedicated to Einstein, Schrödinger's Cat, Alain Aspect, and the Heisenberg Uncertainty Principle.*

I never thought that you'd go beyond
A thought experiment,
Or lose faith in what was
So palpably a truth.

I always thought we were predestined
Not just a roll of dice,
You had to prove me wrong
With oh so practical an aspect.

I wonder now, did what I see
Really exist at all,
Or was it my gaze
That moved us into love?

I taste the fear of genius,
The mindgames, the possibility
That you never would have loved,
The probability that you are loving someone else.

And now, here, locked in my box
Will unseen atoms move the poison along,
And how will I know if I am dead or alive
Without you here to observe?

## The Ratio of Green

*Among the colors, green, which consists in the most moderate action (which, by analogy, one can speak of as a ratio of 1 to 2), is like the octave among musical consonances or like bread among the food that one eats.*
*—Descartes, Treatise on Man*

Among the colors, green, we are the sudden
unexpected but undeniable
razzle of new grass on the fairways, hidden
by a bristling copse of southern pine and maple
behind our rented house. We lie protected
by so many velvet links from the emptying artery
of dark red highway, bloodied with sheer speed.

Our green is sudden but slow, the unlikely
but inevitable offspring of long sleep, revival,
and then blind inches by inches spiraling
through earth's brown leafy curls until the octave
rings openly in air, so its vibration
moves outward as impalpable spheres, as heavens
raised above the players' lowered ken.

Still half-asleep they linger, listening,
as if they merely respected the ninety-degree rule
but, slipped from the embrace of that expanding
music now translated higher, feel
their loss as satisfaction with the green
burgeoning underfoot, and walk away still
slightly dizzy, ears ringing, looking down.

Then we get up together, conjuring loaves
of bread for our breakfast table in plain air.
Analogy makes us the ratio of one
to two: two by virtue of separate stations,
one by virtue of music, a common color
woven like memory through our conversation.
The clear relation of one to two is love.

JoAnne Growney

# *Floating*

My best dream is floating
with legs and arms spread,
paper doll spiraling the wind.

I relax with some people—
ones that don't like me:
my words can't harm;
I won't disappoint.

My best dream is floating
cool-sleek in water, circled by whales—
taking their drift, dropping gently down.

Age resets my evening,
eight o'clock disappears.
Words escape
at the speed of light.

My best dream is floating
at silent breakfast tables with men I don't know
whom I love then and the rest of my life.

Contraptions spin my head,
numerals count my numb days.
My tongue wraps
around comfortable words.

My best dream is floating
from number to number, but between two
are more. Infinity in a moment.

## *Entropy*

She said that entropy
would prove itself,
what did she allude to?
Did she insinuate
her earth had reached its end,
or that, like the universe
it constantly would grow
and never end.

Or did she try to indicate
that man is powerless
against his destiny,
or did she mean to say
that all is chance?

When he suggested
that they should discuss
this cogent problem
over a cup of tea
or something stronger,
she got the cue
and with a smile she said:
Your place or mine?

# H. L. Hix

## From *Orders of Magnitude*

When Thales learned to measure in his head
the surface of earth became a ledger.
The annual Nile flood that gave us mud
gave us maps, calendars imposed on space,
the tablature of number's song. If light
does not seep through a surface, water will.
Burned skin rises, buried bones find the sea.
Pass through one portal into another.
Drive the square-mile county roads in Kansas.
Count the soft woven squares that conceal us.

*

Once a couple of lovers. A couple
of cold friends now. How impotent the hand,
how infinitesimal the abyss
fingertip to fingertip when we stretch,
you prone across the ice, me the scared kid
who fell through. Close enough they would have sparked
once. Close enough to attract iron filings
formed after our former affinity.
Every couple repeats the history
of tangents that once touched but touch no more.

*

The thirty-third time I heard it, your name
became the blaze that cleared the parched prairie,
the fire that opened cones under old growth.
Nothing nurtures like repetition,
or so exposes us as sagittal
sections arranged in rows. No thing repeats,
but all proportions do. Powers of ten,
atom to galaxy. You are to me
as fire to canvas, as tree ring to earth,
as number to a subjugated god.

*

God being gone, love having left, our sole
remaining task is to define complex
numbers. We know they follow daffodil
but anticipate iris and dogwood.
Early blossoms up, final flurries down
figure the square root of negative one.
Something other than integer squeezes
color from frost. Only complex numbers
can tell how far we need-based petals fall,
our very buds beginning our descent.

*

Lord let my lover be no less laissez
than she is fair, engender her, Jesus,
so she wants out of her jeans as often
as I want in them, make her generous
as algebra, who gives more than is asked
of her, who surprises petitioners
with unexpected solutions to cold
nights and thunderstorms, knows y and y naught,
likes the spot x marks (lingers over it),
kneads perfect numbers from bellies and knees.

---

Note from the editors: *Orders of Magnitude* consists of a collection of one-hundred "decimals"—ten-line stanzas with ten syllables per line.

## *Self-Improvement*

Just before she flew off like a swan
to her wealthy parents' summer home,
Bruce's college girlfriend asked him
to improve his expertise at oral sex,
and offered him some technical advice:

Use nothing but his tonguetip
to flick the light switch in his room
on and off a hundred times a day
until he grew fluent at the nuances
of force and latitude.

Imagine him at practice every evening,
more inspired than he ever was at algebra,
beads of sweat sprouting on his brow,
thinking, *thirty-seven, thirty-eight,*
seeing, in the tunnel vision of his mind's eye,
the quadratic equation of her climax
yield to the logic
of his simple math.

Maybe he unscrewed
the bulb from his apartment ceiling
so that passersby would not believe
a giant firefly was pulsing
its electric abdomen in 13 B.

Maybe, as he stood
two inches from the wall,
in darkness, fogging the old plaster
with his breath, he visualized the future
as a mansion standing on the shore
that he was rowing to
with his tongue's exhausted oar.

Of course, the girlfriend dumped him:
met someone, après-ski, who,
using nothing but his nose
could identify the vintage of a Cabernet.

Sometimes we are asked
to get good at something we have
no talent for,
or we excel at something we will never
have the opportunity to prove.

Often we ask ourselves
to make absolute sense
out of what just happens,
and in this way, what we are practicing

is suffering,
which everybody practices,
but strangely few of us
grow graceful in.

The climaxes of suffering are complex,
costly, beautiful, but secret.
Bruce never played the light switch again.

So the avenues we walk down,
full of bodies wearing faces,
are full of hidden talent:
enough to make pianos moan,
sidewalks split,
streetlights deliriously flicker.

Roald Hoffman

## *Why Does Disorder Increase in the Same Direction of Time as That in Which the Universe Expands?*

It has something to do
with looking down the blouse
of the girl painting the boat, tracing
in a second the curve, wanting
to slip a hand between cotton
and her warm skin.

Or seeing a glint of sun
off the window opening across
the bay, calculating the speed
with which the reflection
skims across water.

The girl runs her hand
through her hair, the immemorial
action, this time arrested
as she spots the hummingbird
taking its hovering time
to sample each larkspur blossom.

Or the oil storage tanks
across the water, seeing
them ignite,
silently, the shrapnel
already on its way here.

---

Note from the poet: See S. Hawking, *New Scientist*, July 9, 1987, p 46.

## Jonathan Holden

### *Sex and Mathematics*

*Wovan man nicht sprechan kann,*
*darüber muß man schweigen.*
*—Ludwig Wittgenstein*

Making love we assume
may be defined by the equation
for the hyperbola $y = 1/x$, how
as the denominator
of a warm afternoon in May,
freighted with roses, heavy
with honeysuckle, heavy with the gravity
of one, three,
even five intervals of
unremitting time that
continues on and
on endlessly
like an ear ache
(but a good one) or
interrupted by other
intervals of simply
relenting when simply
to gulp air . . . On certain days
I swear we could see
the asymptote out there
where the curve dies.
It is a view of the sea
napping no more than
half a mile away,
royal blue and peaceful,
a flat line
while the other
becomes so steep
it hugs harder, even harder
the bed-frame,
clings to the opposite wall

like a fly, a kind
of pain, approaching
but not quite
but not
but so
close
oh
s

## X. J. Kennedy

# *Geometry*

They say who play at blindman's buff
  And strive to fathom space
That a straight line drawn long enough
  Regains its starting place
And that two lines laid parallel
  Which never stop nor swerve
At last will meet, for, strange to tell,
  Space throws them both a curve.

Such guesswork lets my hopes abide,
  For though today you spurn
My heart and cast me from your side
  One day I shall return;
And though at present we may go
  Our lonely ways, a tether
Shall bind our paths till time be through
  And we two come together.

Jacqueline Lapidus

## *Several Hypotheses and a Proposition*

nothing's been quite the same with me
since you and I had a falling-out
or should I say throwing since you
tore up all those pictures
of me and I threw you
out of my house    for one thing
I trust myself more and other people
less    for another, we don't write letters
not having you to argue with
alters my inner space    I spend whole nights
meditating with selves I didn't
know I had and wondering    whether we
made each other up
                                        or drawing graphs
on which we appear as two sides
of a right triangle    one upright one
flat    the hypotenuse of course
is the man who came between us
and held us irrevocably
perpendicular    without him we
could have extended ourselves
to infinity    but wherever we end
we always    start    at zero

and whatever we tried, we always got
nowhere    you couldn't love him
and me if we both loved you    I
couldn't love you and him unless
you both loved me    and he couldn't
love both of us no matter what
and unless we both loved him
he couldn't love himself

I don't know what theorem that proves
but I do know whose calculations

determined the result    you
threw him and me together till
our passion became acute    you threw
jealous fits to the point of being
obtuse    then you and I lay
naked in each other's arms
and psychoanalyzed the situation
to the nearest decimal place
                                        finally
you got violent, and that's where
I stopped the vortex and got off

I felt dizzy for a long time
after that    but now the ceiling
and the chairs and the bed have
settled into their proper perspective
and other women to whom I tell the story
say we were all mad

I'm not sure, though    I think
you only offered what you knew
I'd take    and I only accepted
what you wanted me to have
and the man we nearly died of
knew exactly what he was doing and
cancelled out of the equation just in time

one of these days
we'll intersect again

## Counting

Thinking in terms of one
Is easily done—
One room, one bed, one chair,
One person there,
Makes perfect sense; one set
Of wishes can be met,
One coffin filled.

But counting up to two
Is harder to do;
For one must be denied
Before it's tried.

## From *The Cyberiad*

Come, let us hasten to a higher plane,
Where dyads tread the fairy fields of Venn,
Their indices bedecked from one to *n*,
Commingled in an endless Markov chain!

Come, every frustum longs to be a cone,
And every vector dreams of matrices.
Hark to the gentle gradient of the breeze:
It whispers of a more ergodic zone.

In Riemann, Hilbert or in Banach space
Let superscripts and subscripts go their ways.
Our asymptotes no longer out of phase,
We shall encounter, counting, face to face.

I'll grant thee random access to my heart,
Thou'lt tell me all the constants of thy love;
And so we two shall all love's lemmas prove,
And in our bound partition never part.

For what did Cauchy know, or Christoffel,
Or Fourier, or any Boole or Euler,
Wielding their compasses, their pens and rulers,
Of thy supernal sinusoidal spell?

Cancel me not—for what then shall remain?
Abscissas some mantissas, modules, modes,
A root or two, a torus and a node:
The inverse of my verse, a null domain.

Ellipse of bliss, converge, O lips divine!
The product of our scalars is defined!
*Cyberiad* draws nigh, and the skew mind
Cuts capers like a happy haversine.

I see the eigenvalue in thine eye,
I hear the tender tensor in thy sigh.
Bernoulli would have been content to die,
Had he but known such $a^2 \cos 2\Phi$!

Translated from the Polish by Michael Kandel

## Marjorie Maddox

### *Venn Diagrams*

There, stuck in that class,
chalking circles on a board so high your toes ached,
an inch of sock exposed,
all for the sake of subsets, intersection.
That teacher with the tie too bright for day,
wide as your fingers spread

he knew. How even now
worlds swerve in, out, curvilinear,
a trajectory, an extrapolation from that fourth grade:
cowlick, shy wink, lunch box, desk carved with initials,
stacked, bisected, bisected again into lives or

one life divided recursively,
your miracle, you
halved like loaves and fishes.

Mornings you sit, slicing bread,
point in a line between you and . . . .

Through the window the world juxtaposes itself.
Drill perpendicular to concrete.
A jay: coordinate in the grid of an oak.
You part your hair diagonally, unfold the paper.

Those Venn diagrams,
circles with the double cross,
shaded gray, are now.
The overlap: same mouth, nose.

You tilt into different lives without breathing,
love ten people at once/no one.
You eat an apple,
tap your foot to Zeppelin,
fingers to Bach.

Do you see? The circles shift.
Pry your fingers in.
Behind the chalk.
Behind the two dimensions.

## Andrew Marvell

# *The Definition of Love*

My love is of a birth as rare
As 'tis for object strange and high;
It was begotten by Despair
Upon Impossibility.

Magnanimous Despair alone
Could show me so divine a thing
Where feeble Hope could ne'er have flown,
But vainly flapp'd its tinsel wing.

And yet I quickly might arrive
Where my extended soul is fixt,
But Fate does iron wedges drive,
And always crowds itself betwixt.

For Fate with jealous eye does see
Two perfect loves, nor lets them close;
Their union would her ruin be,
And her tyrannic pow'r depose.

And therefore her decrees of steel
Us as the distant poles have plac'd,
(Though love's whole world on us doth wheel)
Not by themselves to be embrac'd;

Unless the giddy heaven fall,
And earth some new convulsion tear;
And, us to join, the world should all
Be cramp'd into a planisphere.

As lines, so loves oblique may well
Themselves in every angle greet;
But ours so truly parallel,
Though infinite, can never meet.

Therefore the love which us doth bind,
But Fate so enviously debars,
Is the conjunction of the mind,
And opposition of the stars.

## *Safety in Numbers*

The enthusiasm with which I repeatedly declare you my one
And only confirms the fact that we are indeed two,
Not one: nor can anything we do ever let us feel three
(And this is no lisp-like alteration: it's four
That's a crowd, not a trinity), and our five
Fingers and toes multiplied leave us at six-

es and sevens where oneness is concerned, although seven
Might help if one was cabalistically inclined, and "one"
Sometimes is. But this "one" hardly means one, it means five
Million and supplies not even an illusion of relevance to us two
And our problems. Our parents, who obviously number four,
Made us, who are two; but who can subtract us from some mythical
three

To leave us as a unity? If only sex were in fact "six"
(Another illusion!) instead of a sly invention of the seven
Dwarves, we two could divide it, have our three and, just as four
Became two, ourselves be reduced to one
—Actually without using our three at all, although getting two
By subtraction seems less dangerous than by division and would also
make five

Available in case we ever decided to try a three-
some. By the way, this afternoon while buying a six-
pack at the Price Chopper as well as a thing or two
For breakfast, I noticed an attractive girl sucking Seven-
Up through an angled and accordioned straw from one
Of those green aluminum containers that will soon litter the four

Corners of the visible world—anyway, this was at five
O'clock, I struck up a conversation with a view to that three-
some, don't be shocked, it's you I love, and one
Way I can prove it is by having you experience the six

Simultaneous delights that require at the very least seven
Sets of hands, mouths, etcetera, anyway more than we two

Can manage alone, and believe me, of the three or four
Women that ever appealed to both of us, I'd bet five
To one this little redhead is likeliest to put you in seven-
th heaven. So I said we'd call tomorrow between three
And four p.m., her number is six three nine oh nine three six.
I think you should call. What do you mean, no? Look, if we can't be
one

By ourselves, I've thought about it and there aren't two
Solutions: we need a third party to . . . No, I'm not a four-
flusher, I'm not suggesting we jump into bed with six
Strangers, only that just as two plus three makes five,
Our oneness is what will result by subtracting our two from three.
Only through multiplicity can unity be found. Remember "We Are
Seven?"

Look, you *are* the one. All I want is for the two
Of us to be happy as the three little pigs, through the four
Seasons, the five ages, the six senses, and of the heavenly spheres all
seven.

---

Note from the poet: Jacques [Roubaud] invented a sestina in which all numbers from 1 to 7 can appear in a strictly regulated order. My poem demonstrates this procedure by using these numbers as end-words.

## Harry Mathews

# *Shakespeare on Love*

Shall I compare thee to a summer's day
And dig deep trenches in thy beauty's field?
Why lov'st thou that which thou receiv'st not gladly,
Bare ruin'd choirs where late the sweet birds sang?
Anon permit the basest clouds to ride
And do what'er thou wilt, swift-footed Time:
Nor Mars his sword, nor war's quick fire, shall burn
Even such a beauty as you master now.
Love's not Time's fool, though rosy lips and cheeks
(When other petty griefs have done their spite,
And heavily) from woe to woe tell oe'r
That Time will come and take my love away;
For thy sweet love remembered such wealth brings
As any she belied beyond compare.

---

Note from the editors: Mathews' sonnet is "constructed" from a permutation of lines, one each from fourteen of Shakespeare's love sonnets. The sonnet numbers, before rearrangement: 2, 3, 18, 19, 29, 30, 33, 55, 64, 73, 90, 104, 116 and 130.

## 3.141592 . . .

*. . . and Cetus chained Andromeda until he was slain by Perseus,*
*who then married her . . .*

In school I was attracted
to irrational numbers
stretched out like variable stars

across an expanding void . . .

They're just trouble Miss MacDougall
said trying to lure me back
to *a* over *b*  something above zero
we could get hold of

solid as carrots and good for our eyes

but I was a born radical
and hunted numbers salty
as Cetus the sea monster
who waylaid maidens in his
starry lair   I wanted to run

in the wrong circles
like the Princess Adromeda
chained unseen in the skies
of my childhood but whose story
and happy ending I believed
with all my absurd heart because
I longed for something like
the square root of love
which I thought would be

before I passed my prime
permanent and easy as pi . . .

## *Distances*

Some distances cannot be crossed; like
Zeno's arrow you can only go halfway at a time:
there remains a remoteness, a shadow thrown
across an almost infinitesimal line:
a separation.
I am usually glad there is a distance between us:
it gives me somewhere to go.
But now, you are 467 miles away
as the crow flies, and I think
That's not a bad number: 4 + 6 makes 10,
the perfect figure, minus 7 makes 3,
the holy trinity or the eternal triangle
neither of which interests me particularly
though I am obsessed by numbers.
I also think, that I am not a crow,
and the actual distance from these shores to you
by shipboard, camelback, Greyhound, underground,
is considerably longer. I can remember
there were times when I could not tell
where I ended or you began

My lost pilgrim
the contours of your body defy distances
and cannot be measured by instruments or statistics:
the distance, for example, between your knee and your ankle
is approximately the distance between
the crow's shadow in the evening
and the soft scent of gardenias;
and the hollow of your absence is wider
than the sound of seashells in September

## The Theory of Absence

The hypothesis: I am tense and so are you.
We neither meet nor separate.

The desired result: We meet in the absence.

The proof: As tension turns people into arcs, we are two arcs.
We neither meet nor separate (the hypothesis)
so, we must be parallel.
If two parallel lines are bisected by a third line
(in this case, the line of tension)
their corresponding angles must be equal (a geometrical theorem).
So we are congruent (because shapes are congruent
when their angles are equal)
and we form a circle (since the sum
of two congruent arcs
is a circle).
Therefore we meet in the absence
(since the circumference of a circle
is the sum of contiguous points
which can each be considered
a point of contact).

Translated from the Arabic by Elizabeth Winslow

Deanna Nikaido

## July 18, 2005

Show me a sign
that factors fear agreeably
over the denominator of my discomfort
so I reduce proportionately
all the weight
that keeps me from solving
this light bearing equation of love.

Bin Ramke

## From *Gravity & Levity*

The heron resolves itself from the gray lake the water
conversely the woman dissolves in sex, her own

in liquefaction but the flesh reforms like wings
unfolded flight like light drips glistens

the setting sun the horizon first
above now below the bird the evening only local

the spinning earth flings its fluid surface
dissolving itself into itself its ecstasy

the need we feel each for each, the falseness
of any world, at all it is a kind of patience

impossible to distinguish from lassitude
it is a kind of hope indistinguishable

from stupidity. I know (of) a man who killed
himself and the woman he was about to marry

killed herself a month later. He wrote a note:
*Until yesterday I had no definite plan to kill myself.*

*I do not understand it myself, but it is not*
*because of a particular event, nor of an explicit matter.*

*Every elliptic curve defined over the rational field*
*is a factor of the Jacobian of a modular function field*

was another note he wrote. (I have his picture
on my desk, a gray parallelogram,

a thin man in black jacket black
tie bifurcating a horizon behind him

the line just above his ears this point
of view this lonely life there is only

a kind of barrenness in the background and a sky
which is a world, of course, plenty.)

This is a bigger world than it was once
it expands an explosion it can't help it it has

nothing to do with us with whether we know or
not whether our theories can be proved

whether or not a mathematician
knew a better class of circles

(he has a name, Taniyama, a Conjecture)
than was ever known before before—

not circles, elliptic curves. Not doughnuts.
Not anything that is nearly, only is, such

a world is hard to imagine, harder to live in,
harder still to leave. A little like love, Dear.

# On Your Imminent Departure: Considering the Relative Importance of Various Motions

Which is more important, the motion of the wind
Forcing every top-heavy reed along the shore
To precisely the same height, or the manipulations
Of the moon moving the white lines of our window slowly
Across the wall of this room?

Or your hand in motion across my back, suggesting
The scarcely noticeable rising of the lake,
The possible empty sky deepening without interruption
In this bed moving relative to the moon
Twenty-five miles away from two minutes ago?

Consider the various subtleties becoming singular
In the eleven motions of our bodies
Pinpointed together here on the night-facing side
Of the earth tilted in summer angle toward the sun
Consistently dragged by the galaxy further out
Into no known direction whatever.

The mind is the only object
That can ever return to this same spot exactly
Over and over.

Beside the calm of the pine trees brushing
Against each other in the dark, how important
Is the hard twist of my refusal to believe in your departure,
You going away alone, westward at one speed
Over forests moving eastward at another, above clouds
Creating southward-flowing shadows touching
Everything left below? Inside this room, how irreversible
Is the steady speed of the inevitable, keeping up,
Always keeping up with us?

Picture the motion of my voice rising now
To ask you this question, leaving itself forever afterward
Motionless in mid-air?

# *Math Is Beautiful and So Are You*

If n is an even number
*then I'll kiss you goodnight right here,*
but if the modulus k is the unique solution,
*I'll take you in my arms for the long night.*

When the properties are constrained as well as incomplete,
*I'll be getting off the train at this stop.*
However, if there is some positive constant,
*then I'll stay on board for a while longer.*

When it says that the supremum deviates from the least zero,
*my heart closes off.*
But if all moments are infinite and you can hear me,
*I will open out for you.*

This sequence satisfies the hypothesis of uniformity,
and because we know that approximation is possible
and that inequality is an embedding factor,
*come, let's try once more.*

# Eryk Salvaggio

## From *Five Poems about Zero*

Because the zed
cannot exist
without memory.
Who could have
no apples?

The apples must
leave traces
of their absence
in your memory—
bright green shining
in the palm of your hand
raised by a delicate elbow
to your mouth.

Zero is a number
of yearning.

In your absence,
I have nothing.
But it's mine.

# The Freezing Point of the Universe

I used to speak in anagrams during sex no wonder
you often left me for girls uninterested in the shift

between "Fawlty Towers" and *flowery twats*. Scold me.
Whenever I think in four-dimensional hypercubic

numerals (1, 16, 81, 256, and so on) you have the right
to demand a simple lunch (pot roast, corn) and tip

me on my head for equilibrium. You're off again I know it,
eyes glazed with dull numbers (although the set

of dull numbers is a null set, go figure). Randomness
steeps in the eye of the beholder; willfulness percolates.

Asked to choose a random number between 10 and 20
you confidently choose 17 like everyone else, a maximally

unremarkable number and here is the catch:
You wish I were one of the Nine Virtuous Women don't you,

the middle pretty sister of the Seven Sisters of Sorrow
and who can blame you. All this talk of radios and "10-codes."

In Shippensburg, Pennsylvania, a 10-45 means "automobile
collision." Elsewhere in the same commonwealth

it stands for "carcass of an unlucky beast." In Maine,
"domestic disturbance." 10-4, Good Buddy. The difference

between the number of pebbles in Newton's calculus
and this four-room house which exhausts the potential

for expansion in the dimensions of width and depth
seems a churlish substitute for the flinty accolades you've been

dealing me lately. And why shouldn't you. Absolute Zero
is where it all begins, the clean slate. Walk out now, you're freezing.

# King Solomon

## From *The Song of Songs*

Threescore are the queens,
fourscore the king's women,
and maidens, maidens without number.

One alone is my dove,
my perfect, my only one,
love of her mother, light
of her mother's eyes.

Every maiden calls her happy,
queens praise her,
and all the king's women:

*"Who is that rising like the morning star,*
*clear as the moon,*
*bright as the blazing sun,*
*daunting as the stars in their courses!"*

Translated from the Hebrew by Ariel Bloch and Chana Bloch

# C. K. Stead

## From *Walking Westward*

October she phoned to say
for her at last it was         over
                                        forgotten
                                        irrelevant
3 nails meant for the heart

might have made use of the new maths
that has a cold beauty
like the beauty of a fiction

as for example that a survey of 19 love affairs showed
        17 were over
         7 were forgotten
         and 13 irrelevant
         but only 2 were all three

9 were over and irrelevant but not forgotten
5 were over and forgotten but not irrelevant

how many that were over were neither irrelevant nor forgotten?

to which a Venn diagram
viz:

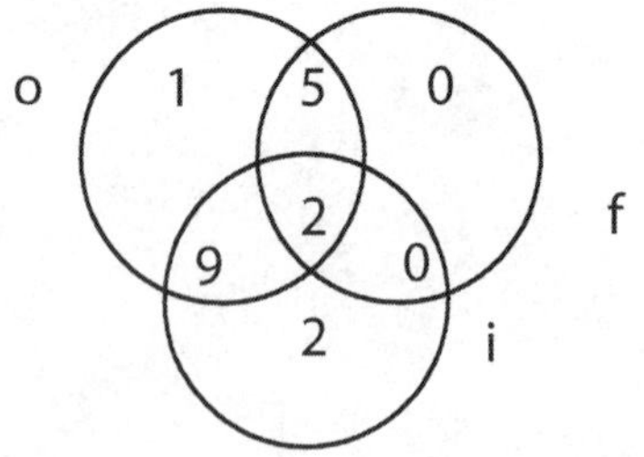


returns the answer 1
(rendered poetically: one only)
                irrefutable

as to say in the language of another dimension
he had explored with her
the caves of generation and the terraces of the stars

## The Meaning of Zero: A Love Poem

*And is where space ends*
*called death or infinity?*
—*Pablo Neruda, The Book of Questions*

A mere eyelid's distance between you and me.

It took us a long time to discover the number zero.

John's brother is afraid to go outside.
He claims he knows
the meaning of zero.

I want to kiss you.

A mathematician once told me you can add infinity
to infinity.

There is a zero vector, which starts and ends
at the same place, its force
and movement impossible
to record with
rays or maps or words.
It intersects yet runs parallel
with all others.

A young man I know
wants me to prove
the zero vector exists.
I tell him I can't,
but nothing in my world
makes sense without it.

## Amy Uyematsu

# Unfinished Calculations

*How true is the line that keeps threading its way through,*
*too thin to cipher and with no start or end*

*when did I first hear a voice that keeps humming in my ear*
*when does a kiss weigh more than a fist, the kiss outlast the scar*

*is forgiveness always less than the magnitude of rage*

*after so many days of walking, what roads does the foot remember*

*how hard must wings push to lift up the body*
*how much rain to drown out the clamor of the mind*

*can anyone measure the diameter of want, the exponential distortion*
*of the glutted or starving eye*

*how many times must the human heart break*

*what is the sound of a singular stone,*
*spiraling into the center of the pond*
*where it ripples the surface in concentric circles,*
*each one drawn into the water's stillness*

*if each baby begins with one billion brain cells, how many*
*electrical signals sent out in the dark will be answered,*
*how many more neurons will die*

*how slim are the odds that one shall paint, another shall dance,*
*whose unexpected orbits keep intersecting mine*

*what is the slope of this spine I lean on, the arc of a lover's arm*
*how soon before the wrinkles on my neck cannot be smoothed with cream*
*why does the distance between us keep shifting*
*what is the meaning of one*

*how many pine needles can sweeten an entire forest*
*how wide does the sky open or close each time someone prays*

*what unproven dimension makes God as close as the wind which keeps*
*stirring my hair, is that why I listen*
*to the rustling of leaves*

*how deep is the garden where this heartroot descends*

## Parabolic Ballad

Along a parabola life like a rocket flies,
Mainly in darkness, now and then on a rainbow,
Red-headed bohemian Gauguin the painter
Started out life as a prosperous stockbroker.
In order to get to the Louvre from Montmartre
He made a detour all through Java, Sumatra,
Tahiti, the Isles of Marquesas.

With levity
He took off in flight from the madness of money,
The cackle of women, the frowst of academies,
Overpowered the force of terrestrial gravity.

The high priests drank their porter and kept up their jabbering:
"Straight lines are shorter, less steep than parabolas.
It's more proper to copy the heavenly mansions."

He rose like a howling rocket, insulting them
With a gale that tore off the tails of their frock-coats.

So he didn't steal into the Louvre by the front door
But on a parabola smashed through the ceiling.

In finding their truths lives vary in daring:
Worms come through holes and bold men on parabolas.

There once was a girl who lived in my neighborhood.
We went to one school, took exams simultaneously.
But I took off with a bang,

I went whizzing
Through the prosperous double-faced stars of Tiflis.
Forgive me for this idiotic parabola.
Cold shoulders in a pitch-dark vestibule . . .
Rigid, erect as a radio antenna rod

Sending its call sign out through the freezing
Dark of the universe, how you rang out to me,
An undoubtable signal, an earthly stand-by
From whom I might get my flight bearings to land by.
The parabola doesn't come to us easily.

Laughing at law with its warnings and paragraphs
Art, love, and history race along recklessly
Over a parabolic trajectory.

He is leaving tonight for Siberia.
Perhaps
A straight line after all is the shorter one actually.

Translated from the Russian by W. H. Auden

## Mathematics: Insects

*for Liz*

*Mathematics possesses not only truth but supreme beauty*
*—a beauty cold and austere, like that of a sculpture.*
*—Bertrand Russell*

Drawn by the window-light, two insects,
contraptions of wire & glass perform
euclidian love. We are safe to assume
they have arrived at these positions
from first principles, *ab ovo*, so to speak,
or at least without the benefit
of the *Kama Sutra*. Nature beggars
the imagination in so many
unexpected ways.

While we are aground
solidly on the bed, leviathans
making token gestures with our tails.

So where is your Euclid now, my love,
whose elegant proposals postulate
the perfect forms by laying on of lines?

# Ronald Wallace

## Chaos Theory

1. *Sensitive Dependence on Initial Conditions*

*For want of a nail the shoe was lost,*
*for want of a shoe the horse was lost,*
and so on to the ultimate loss—a battle,
a world. In other words, the breeze
from this butterfly's golden wings
could fan a tsunami in Indonesia
or send a small chill across the neck
of an old love about to collapse in Kansas
in an alcoholic stupor—her last.
Everything is connected. Blame it on
the butterfly, if you will. Or the gesture
thirty years ago, the glance across
the ninth-grade auditorium floor,
to the girl who would one day be your
lover, then ex-lover, then the wind
that lifts the memory's tsunami,
the mare of the imagination, bolting,
the shoe that claps the nail down on
your always already unending dream.

2. *Love's Discrete Nonlinearity*

No heart's desire is repeatable, or,
therefore, predictable. If a few hungry foxes
gorge on a large population of rabbits,
the population of foxes increases
while that of the rabbits declines,
until some point of equilibrium is passed
and the foxes begin to vanish with
the depleted supply of rabbits, and then
the rabbits multiply, like rabbits. And so on.
The ebb and flow of desire and fulfillment
is a story as old as the world. So,

if I loved you, finally, too much, until
you began to disappear, and I followed,
would you theoretically return to love
repeatedly again? There are forces so small
in our story of foxes and rabbits
no Malthus could ever account for them.
Whole species daily disappear, intractable
as weather. Or think of a continent's
coastlines, their unmeasurable eddies
and whorls: infinite longings inscribed
by finite space and time,
the heart's intricate branchings.

3. *Strange Attractors*

Our vision is simply not large or small enough
to encompass love's fractal geometry.
Who can know the motion of whorl within whorl
entrancing that paradoxical coastline, the changing
habitat of rabbits, the possibility that,
in the clockwork attraction of the solar
system, some heavenly body may not appear
every few million years, to throw all our
calculations asunder? Which says something
for randomness, which has its own hopeful
story. It's just that the patterns of love
and loss are so limitless that chaos
makes its own beautiful picture in which
we are neither (for all our grand needs
and egos) first cause nor unrepeatable.
We are uniquely strange attractors, love's
pendulum point or arc, time's shape or fancy,
in a system with its own logic, be it
the cool elegance of eternity, or
the subatomic matrix of creation and decay.

## Two Tongues in a Tower

*For Harold Loeb*

Back in the iron tower he said, "When I left the tower
the ground shook under foot." "Did the ground-swell shake
any more," she wondered, "than my legs trembled under me?
What had we, then, to fear but nerves teased by a wind?"
"Was it a wind or the earthquake?" he asked; "Do not open the
window.
Unroll the graph. Plot curves. Compare; predict;
record; I have given the numbers." But her eyes, which proved his
numbers
filled with tears as she heard the wet of the night ask:
"Were there voices calling?" Yes, voices of thin fear
sigh for calm, soaring tall to the iron, grim tower:

"This spent of calm is not the calm we want.
Half our dread is lest the tower fall in flood
or fissure." "How *can* you continue your figures?" she asks. "Or *you*,
your tears?" he answers, and quietly placing the casement *against*
the wind, calls out his report: "It is not the flood; but more
drought and wind. Not yet the earthquake," he says clearly,
"but a wind!" and locks the window. She: "They hear the wind,
Statistician, give them statistics." And he? "They ask for bread."
He has lost the key to the larder. The lid of the bread box
is too thickset for tack hammers. He will go on charting predictions.

But the tongued wings' eleventh hour broadcast
they must answer. Too soon now, mutinous and stern
mariners'll come with skeleton key and sledges, abrupt
on this tower to open the larder or with key and oil or with hammers . . .
"To give longshoremen a handout of dry loaves
and salt fish?" he will ask. "Yes. Fish, large
as loaves; loaves, large as baskets," she will answer.
"Shall weepers be fed also?" he'll wonder, as she'll wonder:
"Shall I have, then, more sufficient company of love,"
as when he said (half aloud) to himself; *"Hear that storm."*

## Yang Mu

### *Geometry: River Goddess*

1

She who arrives across the water
Is a goddess. The distinct oars carry the boat along
Almost without a sound,
A chaste, white boat emerging
Between the pavilion and the willows

A goddess is coming, like the chimes of
A bell flitting across the water
Only to disappear in a moment—
Moonlight on her shoulder,
Frost on the prow and the stern.

2

And therefore, there is a lantern
To mark her trail
On the grid of the water:
Every two squares contain a water lily.
She is gliding, and yet she does not seem to come
Any closer, but afar lingering
There. For whom does she tarry on the islet in the middle of the water?[1]

Each time the lantern
Moves into a square, a water lily
Brightens up the moment,
Like her assenting smile.
Weaving in and out of darkness,
The boat leaves a flickering line in its wake
Between the distant pavilion and the nearby willows.
Is she a goddess? She must be
A goddess, moving like an aroused swan on the wing,
Gracefully, a dragon taming the elements, her beauty surpassing
Chrysanthemums in the autumn and pine trees in the spring.[2]

She is coming my way, but she is not any closer.
The slender oars skim the water
Quietly—a vision beyond reach.

3

Even so,
The moon shines on
The young tree, under which the goddess seems standing
Free, her tresses untied. I shall think of a new name
For the tree of lunar splendor.

The boat glides over the coordinates on the grid.
She begins to sing softly, like
A thin flower blooming and closing in the fog,
A fish net slowly sinking,
Willow leaves brushed by the moonlight,
And frost falling. The lyric of the song says:
> I hear the fair lady beckoning me.
> We shall fly away together
> And build a home in the midst of the water . . .[3]

She is coming, but she does not seem to
Get any closer. And as she tarries,
Her trail sparkles intermittently,
Scarcely analyzable—each light in bloom
Is a water lily.

Translated from the Chinese by Michelle Yeh and the author

---

Notes from the poet:
1. From "Goddess of the River Xiang," attributed to Qu Yuan (343?–278 BC).
2. From "Goddess of the River Luo," by Gao Zhi (192–232).
3. From "Lady of the River Xiang," attributed to Qu Yuan.

Part 2

# *Encircling Love*
## *of Family, Nature, Life, and Spirit*

## Harold Lloyd, Student

Do you have the umbrella?
Avez-vous le parapluie?

No, sir, I do not have the umbrella.
Non, monsieur, je n'ai pas le parapluie.

Alice, I have the hippopotamus.
L'hippopotame for you.
Avez-vous le parapluie?

Oui.
Sí.
Yes.

That, which, who, whose.
If the she-lizard is my friend,
then clearly the he-beetle is your friend.
Was it your fault that it rained?
No, the rain was not your fault.
Alice, Alice, it was my fault,
I who study for you
and for this unknowing fly, flowering nightingale of my glasses.

29, 28, 27, 26, 25, 24, 23, 22.
$2\pi r$, $\pi r^2$
and converted itself into the mule, Nebuchadnezzar,
and your soul and mine into a royal bird of paradise.

Fish no longer sing in the Nile
nor does the moon set for the dahlias of the Ganges.

Alice,
why do you love me with that sad crocodile air
and the deep pain of a quadratic equation?

Le printemps pleut sur Les Anges.

The Spring rains over Los Angeles
in that sad hour when the police
are unaware of the suicide of the isosceles triangles,
the melancholy of a Naperian logarithm
and the *facial unibusquibusque.*

In that sad hour when the moon becomes almost equal
to the whole misfortune
of this love of mine multiplied by X
and to the wings of the afternoon that fold
over an acetylene flower
or a bird of gas.

Of this my pure love so delicately idiotic.
Quousque tandem abutere Catilina patientia nostra?

So sweet and deliberately idiotic,
capable of making the square of a circle cry
and obliging that fool, Mr. Nequaqua Schmitt,
to sell at public auction those stars that belong to the river
and those blue eyes that skyscrapers open to me.

Alice, Alice, my love!
Alice, Alice, my nanny!
Follow me by bicycle through air
even though the police may not know astronomy,
the secret police.
Even though the police may not know that a sonnet
consists of two quatrains
and two tercets.

Translated from the Spanish by Mark Strand

## *Dialectic of the Census Takers*

Last night in the dislabor
of falling asleep, I dreamed
my sheep had gone astray
reproducing Fibonacci sums of lambs
which escaped my repose into your wakefulness.

You asked, "The recipe, could I
have the recipe?" Even as my body
made its nightly mends,
I countered, "Formula,
"you do want the *formula?*"

You sulked,
"I thought you were hungry too."
From the wool of every bleating ewe,
a mathematician sprang
Archimedes, Newton, Gauss.

"More," they demanded. "More."
The sheep couldn't keep pace
with their ability to count.
But you insisted,
"I didn't have enough."

In this world, not all of us
can find our own nourishment.
Hunger keeps us awake.
But tonight, dear Friend,
the sheep are somnambulating
from my pillow to yours.
There is nothing to do
and nothing to eat—
nothing, nothing
but sleep.

## From *The Paradiso: Canto XXXIII*

O Light Eternal fixed in Itself alone,
by Itself alone understood, which from Itself
loves and glows, self-knowing and self-known;

that second aureole which shone forth in Thee,
conceived as a reflection of the first—
or which appeared so to my scrutiny—

seemed in Itself of Its own coloration
to be painted with man's image. I fixed my eyes
on that alone in rapturous contemplation.

Like a geometer wholly dedicated
to squaring the circle, but who cannot find,
think as he may, the principle indicated—

so did I study the supernal face.
I yearned to know just how our image merges
into that circle, and how it there finds place;

but mine were not the wings for such a flight.
Yet, as I wished, the truth I wished for came
cleaving my mind in a great flash of light.

Here my powers rest from their high fantasy,
but already I could feel my being turned—
instinct and intellect balanced equally

as in a wheel whose motion nothing jars—
by the Love that moves the Sun and the other stars.

Translated from the Italian by John Ciardi

Carl Andre

## On the Sadness

The door is closed
We are going to die if the moon changes
The sky is blue then we are going to die if the grass is green
We are going to die then we are going to die if the sea is cold
The window is open
We are going to die if the sky is blue if men grow old
Night comes slowly
We are going to die then the sky is blue if the grass is green
The sky is blue if a girl sings
We are going to die if the sun is hot
Morning comes at five o'clock
We are going to die then we are going to die if the sky is blue
then we are going to die
The grass is green if men grow old
We are going to die if a boy runs
The sky is blue if the sea is cold
We are going to die then the grass is green
Fathers go to work
We are going to die if the sky is blue if the grass is green
Mothers mind their children
We are going to die then we are going to die if men grow old
The sky is blue then the sky is blue
We are going to die if a girl sings
The grass is green then we are going to die
We are going to die then the sky is blue if the sky is blue
The moon changes
We are going to die if the sea is cold
The sky is blue if men grow old
We are going to die then we are going to die if the grass is green
The sun is hot
We are going to die if the sky is blue then we are going to die
A boy runs
We are going to die then we are going to die then we are going
to die
The sky is blue if the grass is green
We are going to die if men grow old
A girl sings
We are going to die then we are going to die if the sky is blue
The sea is cold
We are going to die if the grass is green
The sky is blue then we are going to die
We are going to die then the sky is blue
Men grow old
We are going to die if the sky is blue
The grass is green
We are going to die then we are going to die
The sky is blue
We are going to die

Note from the editors: The lines of Carl Andre's poem "On the Sadness" follow a pattern for the factorization of integers from 47 (which corresponds to the first line of the poem) counting downward to 2. Each of the primes between 47 and 2 corresponds to a poet-chosen phrase; for example, 2 corresponds to "We are going to die." To construct each composite line, its corresponding number is expressed, according to the Fundamental Theorem of Arithmetic, as a product of powers of distinct primes in increasing order. Multiplication corresponds to the conjunction "if" and exponentiation to "then." Andre's poem was originally exhibited as art and consisted of words typed (in the arrangement seen on the preceding page) on an 8 ½" × 11" sheet; the poem is presented here in a typewriter font to suggest the poet's original exhibited image.

## *Counting: Gregor Mendel in the Prelacy*

My companions since childhood, these numbers.
My constant counterparts, as lime kilns
steamed on our green hills

and my father grafted to russet knuckles
a golden apple's fingerlings. (The first stalk
six posts from the gate, and the gate
twelve strides from the pond.)

Each winter, I loved the ermine's harmony,
how it stitched over fresh drifts
the parallel pricks of its tracks. And the pale,

symmetrical petals of snow, how they covered
our seventy houses, our eight hundred
yoke of good arable, good meadowland,
our four hundred ninety souls.

*Holy father, do not think that I think of you less*
*when I think of you mathematically.*

Tomorrow, November closes—
and, polished by frost, the church bells
respond with a clarity. Already,

one-fourth of the compost
is eaten by lime, one-third of the belfry
by shadow. How the second hand ticks!
Stay with me, now, as I wind through my first life. . . .

Robin Chapman

# Nonlinear Function

*Def. 1: Not a linear function*

A nonlinear life doesn't turn out
The way you might expect,
More of the same each year,
Straight line from birth to death—
But more like David Copperfield's
Or Horatio Alger's, the story
Of any infant
Where a look or word
Made all the difference
In who they became today.

*Def. 2: One in which $f(x+y)$ does not equal $f(x) + f(y)$*

This is easily enough understood
By any child of divorce—Mom's house
And Dad's house are not the same
As the house with both Mom and Dad before.
Or think of *f* as happiness,
And know that what they had together
Is not what they have now, whatever
The plus or minus sign of once-upon-a time.

*Def. 3: One in which $f(ax)$ does not equal $af(x)$*

This one's obvious to poets, mail carriers,
Preschool teachers—a salary of *a* for each *x*,
Equal pay, is not the same as 500*a* for the CEO
And what's left over for the rest of us *xes*, even though
The nation's average annual income would remain
The same, and some economists would claim
Measures of average income are perfectly adequate
In charting progress in the economy's name.

## God Loves a Curve

God loves a curve!
Oh, yes—He gives straight lines, as in geometry,
Horizontals for measurement,
Perpendiculars for elevators,
Angles, and the like.
Straight lines are for utility,
Delineate the object, but leave me cold.

God loves a curve
And curves give grace and easy beauty:
Hills and human forms,
Marbles, apples, or a rose;
And that winding railroad track—
These all entrance the eye as nothing straight,
When looking back or forward we do see
The rushing symmetry of motion;
Or when heavenward gazing we
Do witness sweeping arcs of flight;
Or waterborne, the billowing tack
Of sails exulting in the breeze.

Yes, God loves a curve!—
And I the moon at night
In scimitar reflection or full-orbed smile.
But most of all I love a curve
In love's full grace,
Caught in your smile!

For which I wait.

## *Practical Application*

He's teaching her arithmetic,
He said it was his mission,
He kissed her once, he kissed her twice and said,
"Now that's addition."
And as he added smack by smack
In silent satisfaction,
She sweetly gave the kisses back and said,
"Now that's subtraction."

Then he kissed her, she kissed him,
Without an explanation,
And both together smiled and said,
"That's multiplication."
Then Dad appeared upon the scene and
Made a quick decision.
He kicked that kid three blocks away
And said, "That's long division!"

Jennifer Crow

## *Mathematics*

He stands, a boy with hands that tremble,
in a courtyard slicked by rain
and blood, and tosses maple seeds
in the air to spiral onto the stones.
Something greasy marks the slates,
as if they too were once alive
and ready to spring up or fall—but any memory
of that has been crushed out of them
and they lie and wait for time
to grind them to dust.

He is the good son, and doesn't look
at the bodies that dangle by bony wrists
from the top of the gate. He is the good
son, and someday he will walk
the battlements of this castle
as its lord. But until that day,
he bows with the rest, and watches,
and hides from sharp knives in the night.
His mother taught him—

The boy squeezes his eyelids tight,
remembering red wells
too deep to bring up tears. He feels
a seed land on his foot
and jumps back. Magpies dart
around the castle's banners
and shriek at the knights
who ride in and ride out.

In a weak moment, the boy counts—
one body, two, a dozen
and he wonders if their ghosts
remember him, or if they

have gone too far to recall
their stolen lives.
This is the mathematics of power:
adding the dead,
dividing the living,
multiplying the sorrows.

# *Fibonacci Numbers*

At a hard wood desk, on a hard
wooden stool, a mathematician counted rings
of monastery bells next door, seeking
God in multiplying numbers. He considered the problem
of ideal fertility: rabbits in fields of buttercups,
daisies, asters, chicory, reproducing at the rate of one
pair per month. The arithmetical series revealed a mean, a golden
number, controlling the growth of leaves, pinecones
tossed in the fire cold winter nights, cauliflower he despised, seeds
that taunted Eve when she delved into the apple
seeking knowledge and gaining children.

I, too, chant the count of hope
each night: one ovum. One sperm. Two eyes. Three-
letter names—Ann or Ian. Five fingers per hand. A perfect
nautilus spiral of an ear. I long for the fertility
of Fibonacci's numbers, that mystical statistical world
where one plus one equals three.

# Flash Cards

In math I was the whiz kid, keeper
of oranges and apples. *What you don't understand,*
*master,* my father said; the faster
I answered, the faster they came.

I could see one bud on the teacher's geranium,
one clear bee sputtering at the wet pane.
The tulip trees always dragged after heavy rain
so I tucked my head as my boots slapped home.

My father put up his feet after work
and relaxed with a highball and *The Life of Lincoln.*
After supper we drilled and I climbed the dark

before sleep, before a thin voice hissed
numbers as I spun on a wheel. I had to guess.
*Ten,* I kept saying, *I'm only ten.*

## My Dog Practices Geometry

I do not understand the poets who tell me
that I should not personify. Every morning
the willow auditions for a new role

outside my bedroom window—today she is
Clytemnestra; yesterday a Southern Belle,
lost in her own melodrama, sinking on her skirts.

Nor do I like the mathematicians who tell me
I cannot say, "The zinnias are counting on their
fingers," or "The dog is practicing her geometry,"

even though every day I watch her using
the yard's big maple as the apex of a triangle
from which she bisects the circumference

of the lawn until she finds the place where
the rabbit has escaped, or the squirrel upped
the ante by climbing into a new Euclidian plane.

She stumbles across the lawn, eyes pulling
her feet along, gaze fixed on a rodent working
the maze of the oak as if it were his own invention,

her feet tangling in the roots of trees, and tripping,
yes, even over themselves, until I go out to assist,
by pointing at the squirrel, and repeating, "There!

There!" But instead of following my outstretched
arm to the crown of the tree, where the animal is
now lounging under a canopy of leaves,

catching its breath, charting its next escape,
she looks to my mouth, eager to read my lips,
confident that I—who can bring her home

from across the field with a word, who
can speak for the willow and the zinnia—
can surely charm a squirrel down from a tree.

## *Moon Compasses*

I stole forth dimly in the dripping pause
Between two downpours to see what there was.
And a masked moon had spread down compass rays
To a cone mountain in the midnight haze,
As if the final estimate were hers;
And as it measured in her calipers,
The mountain stood exalted in its place.
So love will take between the hands a face. . . .

## Geometry Is the Mind of God

A point is that which has no part.
A line is a breadthless length.
A man in his life is a point on a line:
That which has no part on a breadthless length.
The far horizon is a line made of vanishing points,
Near collision of funnelling views,
Flat as a corpse's EKG.
The line to my back
Is a heart attack of granite and ice,
A tumble of similar opposites.
The opposite of a mountain
Is the ocean or the sky
Or an island in the ocean
Or an island in the sky
Or a thorn on the island, growing.
And what regards the reeling firmament
With sympathy?
If the ocean has an island,
If the point has no part,
I'd say it's a green thorn in the heart.

Note from the poet: The title is a quotation from Johannes Kepler. The first two lines are from Euclid's *Elements*.

## Sarah Glaz

# *Love Story*

If I ever write about you—
he said—
it will be a love story
a story about
how much you want to be loved

Father, do you love
your little girl?
I brought you
a soup full
of numbers
formulas chopped to perfection
integrals fried to a crisp

Father, is it algebra
you wish
for supper
or higher mathematics?

I would lie at your feet
looking up
at your busy mouth
waiting for a smile to appear
a few words of approval and praise
one word of love

How long can one wait?

A lifetime if necessary—

even beyond

# *Proportions of the Heart*

In classical flower arrangement,
Masako says, three major stems occur.
The *shin* stands thirty degrees from vertical.
The *soë*, forty-five degrees,
is just three-quarters of the *shin* in height.
The *hikaë*, three-quarters of the *soë*,
points outwards, low, at seventy-five degrees;
most often this one is a flower.

What a classicist I have become,
impelled by the broad hand of revelation,
that is, experience.
Masako's creatures fill our country house
like novel theorems from the *Elements*;
out of fixed proportion, beauty rises
unlike any that I used to summon
in rented rooms from floppy big bouquets.

A single sweep of branch, unflowering,
another upward twist,
and there's the shape of nothing caught in air,
somehow the proper counterpart of one
or two explosive flowers.
Don't be afraid, she says, her fingers hidden
inside the vase, to put more details in,
as long as they don't interrupt the lines.

The heart's most elegant, extravagant
designs arise, I see,
from careful choice and rapid computation.
In half an afternoon, Masako fills
our baskets large and small, and the clear vases.
Two leans from one, and three from one and two,
and suddenly altogether they compose
their ratios to self-sufficiency.

Even the purple brambles in the field,
cut by Masako, fall in whole ellipses,
and twigs repeat their angles on the branch.
So may you and I and our small flower
flourish in the constraints
space and number pose on families;
and make our tracery around the center
of certain loss more beautiful, and sure.

# *The Shape of Desire*

Tracing an airplane's pale trajectory,
you always point, and finish, "Airplane *gone.*"
Waking from dreams about your babysitter's
dark-eyed, clever daughter, you conclude,
"Lulu *gone,*" and hurry to the door's
long windowpane to see her reappear
freshly composed from memory and clouds.
Now you can say the shape of your desire.

Now you believe that each sidereal item
carries a left-hand banner to describe
through curl and dissipation how it was,
that every friend is summoned by a name,
even in parting. You are wrong, and right
about the frail parabolas of love.

Bob Grumman

## *Mathemaku* No. 10

♡<br>
poetry ) *existence*<br>
somewhere, minutely, a widening<br>
existence

## Israel Har

# The Zero

*for Masha, in lasting love*

When nothing remains
trace a small circle
that the place
won't stay empty. So
Muslim mathematicians
sharpened
man's thought. Creator of the universe,
my little circle craving emptiness,
our fathers told us:
a fly
won't bat a lash
without your gaze falling on it—
creator of the universe
alas
your song and singer are shut in the valley of dejection.

Translated from the Hebrew by Gabriel Levin

Jonathan Holden

# *The Departure of an Alphabet*

My father tested as a genius
in mathematics, but not
in hospitals,
where he would become
the model pupil, obedient,
passive. I was teaching
trigonometry but having trouble
deriving the formula
for the cosine of the sum
of two angles, *Alpha* plus *Beta*.
Could he help? I knew
he needed stimulation.
Mathematics was like sex for him, it
energized him. You could do it
anywhere. Even here,
I thought. But
as he hunched on the bed's edge,
his twig-thin freckled ankles
dangling from pajamas,
and began madly to cover the back
of the envelope I'd handed him
with *Alphas* and *Betas*,
I saw that he was having trouble
with his signed numbers and
his polynomial functions missed
the lines. They sagged, drooping off the edge.

After he'd died,
when I went through the attic, I found
a dusty liquor box packed high
with yellow legal pads
crowded with his algebra,
each concatenation
executed with a fountain pen
at breakneck speed

like an Olympic skier banking
through a gate, then at the margin
digging in to begin the next
concatenation, cutting
back to reach the next
parenthesis, *slant, slant;*
his mind outraced his hand;
the hand probably had trouble
keeping up;
he'd been trying
to solve Fermat's Last Theorem.
And here is how the world will end.

One day, at, say 10 A.M.
there is no more letter E.
That vowel has departed.
We try to read TH NW YORK TIMS.
Other letters are leaving. We've lost
another vowel. The letter A
has bought the farm. Cn w think?
Oops, there goes U. It's confsing.
Now O. Hw cn w cry
r ?

## Addition [1]

$7 \times 7$ + love =
An amount
Infinitely above:
$7 \times 7$ – love.

## Solomon Ibn Gabirol

### From *The Kingly Crown: The Praises of God*

Thou art One, the beginning of all computation, the base of all
construction.
Thou art One, and in the mystery of Thy Oneness the wise of heart are
astonished, for they know not what it is.
Thou art One, and Thy Oneness neither diminishes nor increases, neither
lacks nor exceeds.
Thou art One, but not as the One that is counted or attained, for number
and change cannot reach Thee, nor attribute, nor form.
Thou art One, but my mind is too feeble to set Thee a law or a limit, and
therefore I say: "I will take heed to my ways, that I sin not with
my tongue."
Thou art One, and Thou are exalted high above abasement and falling—
not like a man, who falls when he is alone.

Translated from the Hebrew by Bernard Lewis

## James Kirkup

# *How to Find the Longest Distance Between Two Points*

From eye to object no straight line is drawn,
Though love's quick pole directly kisses pole.
The luckless aeronaut feels earth and moon
Curve endlessly below, above the soul
His thought imagines, engineers in space.
The crewless captain feels high heaven swoon
In elliptic oceans, where deserts roll.
The lost explorer sees the mountain raise
A cryptic summit where no valley is.

Pacing the infinite's familiar place
Where things on things revolve, yet hover still,
And race into star-screwed oblivion,
The longest distance from that point we found
Shortest to this, where nothing is defined.

## Geometry of Moods

Concentrical, the universe and I
Rotated on God's crystal axletree,

So perfectly adjusted in suspense
I could not mark our split circumference,

But sphere in sphere, devotion in devotion,
Was a thing of folding air, a windy motion.

My spinal pole, tipped with a globe of light,
Stretched long as time into the infinite,

And when I turned to look upon the face
Of love's incarnate form, tremendous space

Inclined upon her mouth, the natural nest
Where, throbbing like a bird, singed thought might rest.

Concentrical, the universe and I
Rotated on God's crystal axletree,

I core of the world, a bead in a ball of glass
So pure that only Nothing could be less.

Oh the earth ensphered me, liberal and warm,
When the curve of heaven was her sleeping arm.

Now cubical upon a fractured pole
It creaks, scraping the circle of my soul.

## The Comet

Near from infinity I came
    Drawn to your strong, unmoving light
By some ascendance of its flame
    That charms the planets through their night.
The distance melts, my spirit thaws,
    Sublimes, and in your radiance flies
Soon, by the old, unchanging laws,
    An exhalation through the skies.
Sweet perihelion! May we touch,
    Our auras intermingle? No,
The impulse of my flight too much,
    I must again to darkness go;
While you may stand, and watch my face
    Dwindle through trans-Plutonian space.

Israel Lewis

## *I Find My Faith in the Flatness of Space*

Adherent of convergence, I find solace
in the cold precision of planets and moons
circling their hosts in closed ellipses.

Divergence discomfits me, as from the swoop
of parabolae, the twinned trajectories of hyperbolae.
I am impatient for the tangency of asymptotes

uneasily tolerate the curvature of space, suction
of gravity. Cusps of space-time prick me. Galaxies fly
away in inexorable expansion. I take offense.

## From *Sundry Christian Passions: Square Poem in Honor of Elizabeth I*

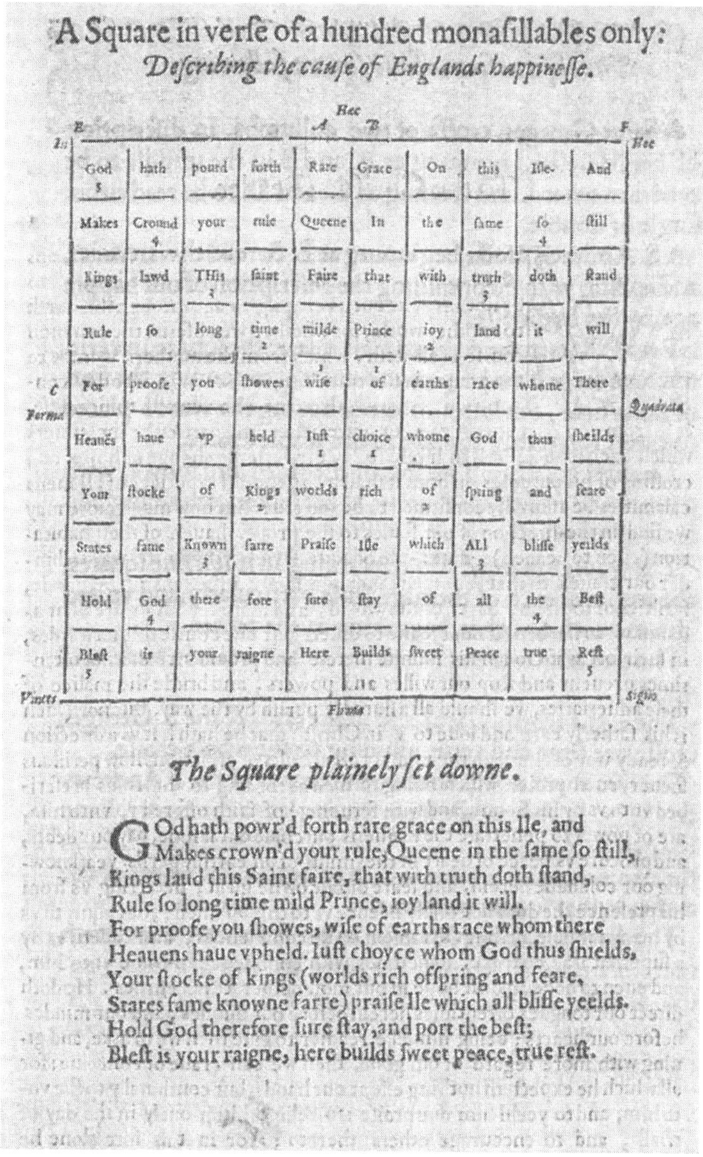

### A Square in verſe of a hundred monaſillables only:

*Deſcribing the cauſe of Englands happineſſe.*

Hec

E A B F

In | Hec

| | | | | | | | | | |
|---|---|---|---|---|---|---|---|---|---|
| God 5 | hath | pourd | forth | Rare | Grace | On | this | Iſle- | And |
| Makes | Crownd 4 | your | rule | Queene | In | the | ſame | ſo 4 | ſtill |
| Kings | lawd | THis 3 | ſaint | Faire | that | with | truth 3 | doth | ſtand |
| Rule | ſo | long | time 2 | milde | Prince | ioy 2 | land | it | will |
| For | proofe | you | ſhowes | 1 wiſe | 1 of | earths | race | whome | There |
| Heauẽs | haue | vp | held | Iuſt 1 | choice 1 | whome | God | thus | ſheilds |
| Your | ſtocke | of | Kings 2 | worlds | rich | of 2 | ſpring | and | feare |
| States | fame | Known 3 | farre | Praiſe | Iſle | which | ALl 3 | bliſſe | yeilds |
| Hold | God 4 | there | fore | ſure | ſtay | of | all | the 4 | Beſt |
| Bleſt 5 | is | your | raigne | Here | Builds | ſweet | Peace | true | Reſt 5 |

C Forma | Quadrata

Vnitis | Sigio

Fluxit

### *The Square plainely ſet downe.*

GOd hath powr'd forth rare grace on this Ile, and
Makes crown'd your rule, Queene in the ſame ſo ſtill,
Kings laud this Saint faire, that with truth doth ſtand,
Rule ſo long time mild Prince, ioy land it will.
For proofe you ſhowes, wiſe of earths race whom there
Heauens haue vpheld. Iuſt choyce whom God thus ſhields,
Your ſtocke of kings (worlds rich offſpring and feare,
States fame knowne farre) praiſe Ile which all bliſſe yeelds.
Hold God therefore ſure ſtay, and port the beſt;
Bleſt is your raigne, here builds ſweet peace, true reſt.

Note from the editors: The ancient square poem on the preceding page has ten lines of ten syllables each. It was published in 1597 and dedicated to Queen Elizabeth I. The square can be read as a poem row by row, but it also includes other geometrically-shaped poems within the original square: a St. George's cross, a St. Andrew's cross, two pillars, and five "squares within squares." For example, the St. Andrew's cross is made by reading downward along the main diagonals, first starting at E, and then starting at F. The Latin mottos that surround the square read: *Haec Quadrata Firma Forma* (This Square is a Strong Form) and *In Hoc Signo Vinces* (In This Sign You Will Conquer). Additional commentary is available in *Petrarch and the English Sonnet Sequences*, by Thomas P. Roche, Jr., AMS Press, 1989, and *Material Poetry of the Renaissance/The Renaissance of Material Poetry*, ed. Roland Greene, Harvard Library Bulletin, #3, 1992.

## Outwitted

He drew a circle that shut me out—
    Heretic, a rebel, a thing to flout.
But Love and I had the wit to win:
    We drew a circle that took him in!

## Wilmer Mills

# *An Equation for My Children*

It may be esoteric and perverse
That I consult Pythagoras to hear
A music tuning in the universe.
My interest in his math of star and sphere
Has triggered theorems too far-fetched to solve.
They don't add up. But if I rack and toil
More in ether than a mortal coil,
It is to comprehend how you revolve,
By formulas of orbit, ellipse, and ring.

Dear son and daughter, if I seem to range
It is to chart the numbers spiraling
Between my life and yours until the strange
And seamless beauty of equations click
Solutions for the heart's arithmetic.

## Geese

My math teacher Epstein
liked to call me to the blackboard.
He said that my head was good only for hats,
and that a bird with brains like mine
would fly backwards.
He sent me to tend the geese.

Now, at a distance of years from his sentence,
when I sit under the palm tree
with my three beautiful geese,
I think that math teacher of mine was farsighted.
He was right,

because nothing makes me happier
than to watch them now
falling upon bread crumbs,
joyful tails wagging,
or freezing for a moment
under beads of water
when I spray them
with a hose,
holding their heads erect,
bodies stretched back
as if remembering faraway lakes.

Since then my math teacher has died,
together with the math problems
I could never solve.
I like hats
and always at evening
when the birds return to the tree
I look for the one flying backwards.

Translated from the Hebrew by Lisa Katz

## *Math*

My new father doesn't like it either.
He loosens his tie, unbuttons
three buttons at the top of his shirt.
Maybe he'll just have one cigarette—
he taps the pack against his arm,
leaves the others waiting in even rows.
*Like this,* he says, lighting up,
and crosses off a number in the tens column,
adds a tiny figure at the top.

When it's my turn
I swing my legs over the edge of my bed,
scuffing my slippers.
Paper thins under the eraser.
One number shoulders another
as he discovers I can't carry.

He breathes out slowly.
A perfect smoke ring
hangs over my bed.

Kyoko Mori

## *Barbie Says Math Is Hard*

As a boy, I'd still have asked
why Jack must spend exactly
two dollars at the corner store.
Give him a coin purse is as
good an answer as five apples
and two oranges. Also: would
he bake the apples into pies
or cobblers, save the orange peel
in glass jars to spice up his
tea or cake? If his father
paints their house with Mr. Jones,
which man will take the peaks and
why? Would the raspberry beetles
swarm over wet paint? Why is
Mr. Jones slower than his
neighbor? If x equals y,
is it like putting apples into
cole slaw, the way a tomato
is really a fruit? None of my
dolls talked or grew hair. In
third grade, Satsuki and I
traded our Barbies' limbs so
mine could flex her left biceps
while hers sat cross-legged
raising one stiff arm
like a weapon. If Satsuki has
daughters, she might remember
the grasshoppers we caught,
how we cupped two hands together
into crooked globes to
hear them rattling inside like
a small motor. She would tell
her daughters: Yes, math was hard,
but not because we were girls.

## Child Margaret

The child Margaret begins to write numbers on a Saturday morning, the
first numbers formed under her wishing child fingers.
All the numbers come well-born, shaped in figures assertive for a frieze
in a child's room.
Both 1 and 7 are straightforward, military, filled with lunge and
attack, erect in shoulder-straps.
The 6 and 9 salute as dancing sisters, elder and younger, and 2 is a
trapeze actor swinging to handclaps.
All the numbers are well-born, only 3 has a hump on its back and 8 is
knock-kneed.
The child Margaret kisses all once and gives two kisses to 3 and 8.
(Each number is a brand-new rag doll.... O in the wishing fingers...
millions of rag dolls, millions and millions of new rag dolls!!)

## *Smart*

My dad gave me one dollar bill
'Cause I'm his smartest son,
And I swapped it for two shiny quarters
'Cause two is more than one!

And then I took the quarters
And traded them to Lou
For three dimes—I guess he don't know
That three is more than two!

Just then, along came old blind Bates
And just 'cause he can't see
He gave me four nickels for my three dimes,
And four is more than three!

And I took the nickels to Hiram Coombs
Down at the seed-feed store,
And the fool gave me five pennies for them,
And five is more than four!

And then I went and showed my dad,
And he got red in the cheeks
And closed his eyes and shook his head—
Too proud of me to speak!

Young Smith

## *She Considers the Dimensions of Her Soul*

The shape of her soul is a square.
She knows this to be the case
because she often feels its corners
pressing sharp against the bone
just under her shoulder blades
and across the wings of her hips.
At one time, when she was younger,
she had hoped that it might be a cube,
but the years have worked to dispel
this illusion of space, so that now
she understands: it is a simple plane,
a shape with surface, but no volume—
a window without a building, an eye
without a mind.
Of course, this square
does not appear on x-rays, and often,
weeks may pass when she forgets
that it exists. When she does think
to consider its purpose in her life,
she can say only that it aches with
a single mystery, for whose answer
she has long ago given up the search—
since its question is a word whose name
can never quite be asked. This yearning,
she has concluded, is the only function
of the square, repeated again and again
in each of its four matching angles,
until, with time, she is persuaded
anew that what it frames has no
interest in ever making her happy.

Elizabeth Anne Socolow

## *Asymptotes*

I can still see the curve I first traced from the equation,
the points plotted, the line drawn with the finest pencil
and the astonishing news: the curve will approach the axis
forever and never actually touch, though the pencil
being real not ideal might be too thick to represent
the exquisite almost of that never.
It seemed a clear metaphor
for the lovers in the highest level of Dante's Purgatorio,
the ones who from an excess of what is to be hoped for
landed in each other's arms, adulterers never able
to consummate their passion, never able to be removed from Hell.
But also, how it might be to reach for knowledge
and have it always elude us, making us yearn
for a summing sort of God. Nothing personal.

I have loved the idea of pattern existing
even in the almost random choices of light passing through a grid,
trying to decide whether to bounce or flow,
like choosing the waltz or the lindy or going to the masked ball as
oneself.
In the days of the beautiful aquamarine graph paper,
I never dreamed the curve would be about laughter, the sharing of the
jig,
and the gig, and its being up always looming. That is how it is now,
asymptotes: moving toward death and God without welcome of me
as if they were the marriage, the dog, the house, the sense of
coming home to someone else that has eluded me.

When clouds are overhead, I can feel the sky moving toward
the earth in a slow, gradual curve—I alone in a boat or bed
wrangling with the dread and panic of what comes lurching,
too sudden and fierce for muscle,
and sheets and rudder.

# A. E. Stallings

## *Sine Qua Non*

Your absence, father, is nothing. It is naught—
The factor by which nothing will multiply,
The gap of a dropped stitch, the needle's eye
Weeping its black thread. It is the spot
Blindly spreading behind the looking glass.
It is the startled silences that come
When the refrigerator stops its hum,
And crickets pause to let the winter pass.

Your absence, father, is nothing—for it is
Omega's long last O, memory's elision,
The fraction of impossible division,
The element I move through, emptiness,
The void stars hang in, the interstice of lace,
The zero that still holds the sum in place.

## Mark Strand

# The New Poetry Handbook

*For Greg Orr and Greg Simon*

1 If a man understands a poem,
he shall have troubles.

2 If a man lives with a poem,
he shall die lonely.

3 If a man lives with two poems,
he shall be unfaithful to one.

4 If a man conceives of a poem,
he shall have one less child.

5 If a man conceives of two poems,
he shall have two children less.

6 If a man wears a crown on his head as he writes,
he shall be found out.

7 If a man wears no crown on his head as he writes,
he shall deceive no one but himself.

8 If a man gets angry at a poem,
he shall be scorned by men.

9 If a man continues to be angry at a poem,
he shall be scorned by women.

10 If a man publicly denounces poetry,
his shoes will fill with urine.

11 If a man gives up poetry for power,
he shall have lots of power.

12 If a man brags about his poems,
he shall be loved by fools.

13 If a man brags about his poems and loves fools,
he shall write no more.

14 If a man craves attention because of his poems,
he shall be like a jackass in moonlight.

15 If a man writes a poem and praises the poem of a fellow,
he shall have a beautiful mistress.

16 If a man writes a poem and praises the poem of a fellow overly,
he shall drive his mistress away.

17 If a man claims the poem of another,
his heart shall double in size.

18 If a man lets his poems go naked,
he shall fear death.

19 If a man fears death,
he shall be saved by his poems.

20 If a man does not fear death,
he may or may not be saved by his poems.

21 If a man finishes a poem,
he shall bathe in the blank wake of his passion and be kissed
by white paper.

# True North 1

The Lover Wishes to Be of Unending

secret service to the beloved. When the beloved is God,
this is hard—
except for the atheist.

A magnetic compass,
even without interference from nearby
masses of iron—the steel frame of a building—doesn't point

to the North Pole; it points
somewhere
in Hudson's Bay.

To find, or to reckon
the direction true North, you need to get help
from the spinning earth.

A lover who eschews force—*anyone*
*coming forward to speak*
*is using force*—

can only
stand waiting, here
on earth, where

there *are* no
straight
lines.

Stephanie Strickland

## *Who Counts, Counts*

Baby and you
—and me,
we will make three,

but baby-and-me
are different: we're two-
who-are-one.

So, together, five—or we *were*, when
I-was-two-in-one,
but

wishing, it was so hot
that summer, I was wishing
we were two.

You and me, we've been two
who were one as well, but nobody thinks
that's the same, or

a problem. How
many of us were there *really*,
when

I-was-two-who-were-one? Was it
five: us-two + we-three?
Or three?

Or two.
You said, "If it came
—God forbid—to that, well then,

just
two." You meant, should it come,
Godsent, to some crux,

should push
come to knife,
just

Baby and you.

## John Updike

### From *Midpoint: The Play of Memory*

Father, as old as you when I was four,
I feel the restlessness of nearing death
But lack your manic passion to endure,
Your Stoic fortitude and Christian faith.
Remember, at the blackboard, factoring?
My life at midpoint seems a string of terms
In which an error clamps the hidden spring
Of resolution canceling confirms.
Topheavy Dutchmen sundered from the sea,
Bewitched by money, believing in riddles
Syrian vagrants propagated, we
Incline to live by what the world belittles.
    God screws the lukewarm, slays the heart that faints,
    And saves His deepest silence for His saints.

## *Loyal, My Heart . . .*

Loyal, my heart gains merit in shadows.
I'd bring it out to the day, as one might fetch
a tongue of fire out of smallest purgatory into light;
and as I hear it beat its prison, I drown
and sink in the remorseful tenderness of a father
who feels his blind child pulse in his arms.

Loyal, my heart gains merit in shadows.
Pleasure, love, sorrow . . . all offends it,
all stimulates its cruel logarithmic run,
its avid tides and eternal surf.

Loyal, my heart gains merit in shadows.
It's miter and valve . . . I'd tear it out
to carry it in triumph to meet the day,
the stole of violets draped on the shoulders of Dawn,
the purple sash of sunsets,
heavenly bodies, the jovial perimeter of women.

Loyal, my heart gains merit in shadows.
From a sharp mountain peak I'll throw it,
a bleeding disk into the blazing sun.
So will I uproot the cancer of my hard fatigue,
and become impassive from East to West,
with a depraved smile to witness
the ineptitudes of inept culture,
and my heart will hold the flame taken
from the symphonic conflagration of the celestial sphere.

Translated from the Spanish by Enriqueta Carrington

# The Lake Swan, the Tom

The lake swan, the tom,
Swimming love's vector
Diagonally across, connects
For her, unassumingly
With his neck's curve,
The mountain and the deep,
Wet black underneath.

How when she comes upon
The same scene later
And, unable to locate
Her eye's companion,
Cries uncompromisingly
For his clear, white
Figure again to appear.

Part 3

# *Unbounded Love for Mathematics and Mathematicians*

# To Divine Proportion

To you, amazing discipline,
ratio: source of beauty without flaw,
revering the rich life cloistered within
the armored confines of your sacred law.

To you, the retina's beatific jail,
golden section, squaring up to scale,
mysterious wellspring of moderation
that gives rise to harmonious Creation.

To you, ocean of angulated dreams,
flower of the five regular forms,
blue dodecahedron, sonorous arc.

You spark a kindled compass with your wings.
A clear sphere is the song your theorem sings.
To you, divine proportion, gold from dark.

Translated from the Spanish by Carolyn L. Tipton

## My Number

*I'm linked with the fate of the world's disasters and only*
*have a little freedom to live or die.*
*—Vîteslav Nezval*

My number is small. A hundred pounds of water,
A quart of salt. Her digit is a garment.

I wear her like a shadow. We judge each other,
My number and I. She is the title. The license.

The cash drawer. My random number.
She protects me from myself. She desires me.

She says she's only one of thirty million species.
She wishes she were more than anecdotal evidence.

Being human she can erect elaborate scaffolding
To protect her emotions, can make an excuse of obvious

Dramatic proportions. My number is inconsequential
With dreams of glory. She spends three or four days each year

just opening her mail. Do you know how many animals
Will be given lethal injections while you read this poem.

Five billion people = half a billion empty bellies.
If there is a god, why can't that god be smaller than my number,

Tiny, soft-spoken so she'd have to pay attention.
My number is a female impersonator (she has multiple meanings).

Her shape is misleading. The further she is from unity, the more deeply
Involved with the world. Like the winds and the grasses, she wears
herself down.

She lies under hot flags of lilies, sings like a bee.
She gets so lonely she recites for her cat. She makes her face up

Like a death mask. She hangs her dresses on the clothesline outside.
Together we dance—my number and her best dresses.

# Jakob Bernoulli

## From *Treatise on Infinite Series*

Even as the finite encloses an infinite series
And in the unlimited limits appear,
So the soul of immensity dwells in minutia
And in narrowest limits no limits inhere.
What joy to discern the minute in infinity!
The vast to perceive in the small, what divinity!

Translated from the Latin by Helen M. Walker

# Bhaskaracharya

## From *Lilavati*

Whilst making love a necklace broke.
A row of pearls mislaid.
One sixth fell to the floor,
One fifth upon the bed;
The young woman saved one third of them;
One tenth were caught by her lover.
If six pearls remained upon the string
How many pearls were there altogether?

Originally written in Sanskrit; translated from the French by David Bellos

Ray Bobo

## *Give Me an Epsilon and I Will Treat It Well*

| | |
|---|---|
| $x = \emptyset$ | Found alone and wanting,<br>I now languish unbegun; |
| $\emptyset \to \{\emptyset\}$ | Oh, to have your arms about me,<br>Be transformed into the one. |
| $\delta = \{\epsilon\}$ | Closeness for our angels<br>Is returned with an embrace; |
| $x + y \to \infty$ | So that by incorporation,<br>We can soar to loving space. |
| $\partial \int f = f$ | In meaning full, |
| $\int \partial f = f + C$ | In asking all, Amen. |

# Where the Kissing Never Stops

Before curves kiss
they have to get acquainted
discuss their known
and unknown functions
settle a discrete problem or two
decide on a common border

They strive to make each other
equal zero
to reach that point
at which they will reduce to lines
and kiss

They linger
at the intersection of multiplicity
although the relationship has been clarified
the performance rated
the equation ended

## From *The Hunting of the Snark: The Beaver's Lesson*

The Beaver brought paper, portfolio, pens,
And ink in unfailing supplies:
While strange creepy creatures came out of their dens,
And watched them with wondering eyes.

So engrossed was the Butcher, he heeded them not,
As he wrote with a pen in each hand,
And explained all the while in a popular style
Which the Beaver could well understand.

"Taking Three as the subject to reason about—
A convenient number to state—
We add Seven, and Ten, and then multiply out
By One Thousand diminished by Eight.

"The result we proceed to divide, as you see,
By Nine Hundred and Ninety and Two:
Then subtract Seventeen, and the answer must be
Exactly and perfectly true.

"The method employed I would gladly explain,
While I have it so clear in my head,
If I had but the time and you had but the brain—
But much yet remains to be said.

"In one moment I've seen what has hitherto been
Enveloped in absolute mystery,
And without extra charge I will give you at large
A Lesson in Natural History."

In his genial way he proceeded to say
(Forgetting all laws of propriety,
And that giving instruction, without introduction,
Would have caused quite a thrill in Society),

"As to temper the Jubjub's a desperate bird.
Since it lives in perpetual passion:
Its taste in costume is entirely absurd—
It is ages ahead of the fashion:

"But it knows any friend it has met once before:
It never will look at a bribe:
And in charity-meetings it stands at the door,
And collects—though it does not subscribe.

"Its flavour when cooked is more exquisite far
Than mutton, or oysters, or eggs:
(Some think it keeps best in an ivory jar,
And some, in mahogany kegs:)

"You boil it in sawdust: you salt it in glue:
You condense it with locusts and tape:
Still keeping one principal object in view—
To preserve its symmetrical shape."

The Butcher would gladly have talked till next day,
But he felt that the Lesson must end,
And he wept with delight in attempting to say
He considered the Beaver his friend:

While the Beaver confessed, with affectionate looks
More eloquent even than tears,
It had learned in ten minutes far more than all books
Would have taught it in seventy years.

They returned hand-in-hand, and the Bellman, unmanned
(For a moment) with noble emotion,
Said "This amply repays all the wearisome days
We have spent on the billowy ocean!"

Such friends, as the Beaver and Butcher became,
Have seldom if ever been known;
In winter or summer, 'twas always the same—
You could never meet either alone.

And when quarrels arose—as one frequently finds
Quarrels will, spite of every endeavour—
The song of the Jubjub recurred to their minds,
And cemented their friendship for ever!

## *Mathematics*

*And the angel of numbers*
*is flying*
*from 1 to 2...*
*—Rafael Alberti*

The numbers
on our childhood slates
form ranks
when the angel with the integral
enters the room

He comes with $e^{-x}$
with $\sqrt[\pi]{n}$
his tunic
is covered
with the snow of infinity

Angel of intuition
with your letters
we illumine
n-dimensional spaces
in the universe

Gone
the Euclidean tranquility
of the world

Translated from the German by Ewald Osers

## *Scared and the Intermediate Value Theorem*

On a connected surface
one can always, quite smoothly and without abandoning
get from any one point to any other.
A man's body is connected.
And I've been at hands, lips, shoulders.
So how come there's that discontinuity at the waistline?
How come, around there, Zeno whispers "halfway?"
How come that waistline is throbbing with infinity?
And my hand and heart throbbing
with zero?

Marion Deutsche Cohen

## *What Drove Me into Math*

What drove me into math
was not Fermat's Last.
I preferred the factoring of the difference of two squares.
And Cantor's stretched-out one-dimensional lace.
Also, the center of a circle is inside the circle.

What drove me into math
was not the Mystery of the Unknown
but the mystery of the known.

Other early influences:
the point of light just happening to coincide with the only visible
corner of our livingroom
those dark-red shapes when you close your eyes tight
and that spot, that nightmare
of many bloody colors.

# Numbers

I like the generosity of numbers.
The way, for example,
they are willing to count
anything or anyone:
two pickles, one door to the room,
eight dancers dressed as swans.

I like the domesticity of addition—
*add two cups of milk and stir*—
the sense of plenty: six plums
on the ground, three more
falling from the tree.

And multiplication's school
of fish times fish,
whose silver bodies breed
beneath the shadow
of a boat.

Even subtraction is never loss,
just addition somewhere else:
five sparrows take away two,
the two in someone else's
garden now.

There's an amplitude to long division,
as it opens Chinese take-out
box by paper box,
inside every folded cookie
a new fortune.

And I never fail to be surprised
by the gift of an odd remainder,
footloose at the end:

forty-seven divided by eleven equal four,
with three remaining.

Three boys beyond their mothers' call,
two Italians off to the sea,
one sock that isn't anywhere you look.

## *Mandelbrot Set*

*song lyrics*

Pathological monsters! cried the terrified mathematician
Every one of them is a splinter in my eye
I hate the Peano Space and the Koch Curve
I fear the Cantor Ternary Set
And the Sierpinski Gasket makes me want to cry
And a million miles away a butterfly flapped its wings
On a cold November day a man named Benoit Mandelbrot was born

His disdain for pure mathematics and his unique geometrical insights
Left him well equipped to face those demons down
He saw that infinite complexity could be described by simple rules
He used his giant brain to turn the game around
And he looked below the storm and saw a vision in his head
A bulbous pointy form
He picked his pencil up and he wrote his secret down

Take a point called Z in the complex plane
Let Z1 be Z squared plus C
And Z2 is Z1 squared plus C
And Z3 is Z2 squared plus C and so on
If the series of Z's should always stay
Close to Z and never trend away
That point is in the Mandelbrot Set

Mandelbrot Set you're a Rorschach Test on fire
You're a day-glo pterodactyl
You're a heart-shaped box of springs and wire
You're one badass fucking fractal
And you're just in time to save the day
Sweeping all our fears away
You can change the world in a tiny way

Mandelbrot's in heaven, at least he will be when he's dead
Right now he's still alive and teaching math at Yale

He gave us order out of chaos, he gave us hope where there was none
And his geometry succeeds where others fail
If you ever lose your way, a butterfly will flap its wings
From a million miles away, a little miracle will come to take you home

Just take a point called Z in the complex plane
Let Z1 be Z squared plus C
And Z2 is Z1 squared plus C
And Z3 is Z2 squared plus C and so on
If the series of Z's should always stay
Close to Z and never trend away
That point is in the Mandelbrot Set
Mandelbrot Set you're a Rorschach Test on fire
You're a day-glo pterodactyl
You're a heart-shaped box of springs and wire
You're one badass fucking fractal
And you're just in time to save the day
Sweeping all our fears away
You can change the world in a tiny way
And you're just in time to save the day
Sweeping all our fears away
You can change the world in a tiny way
Go on change the world in a tiny way
Come on change the world in a tiny way

---

Note from the editors: The Mandelbrot set 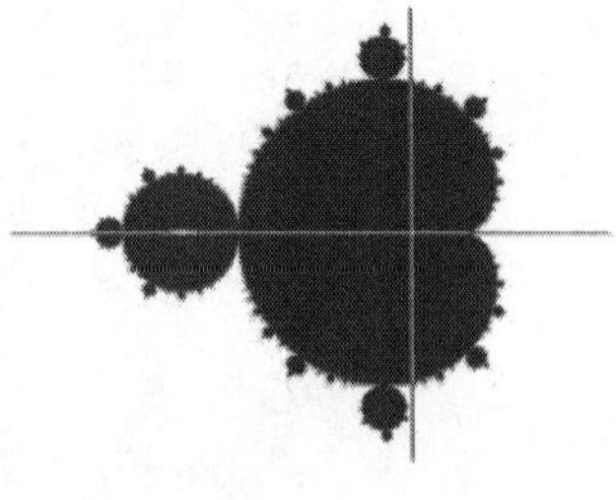

Keith Allen Daniels

## *Mathrimony*

Calculating with exactitude
the definitive parameters
of her connubial existence,
the mathematical genius
considers for the nth time
the exquisite irony of marrying
a man whose "math anxiety"
was nearly phobic in its intensity.

Perhaps his sexuality index, s,
appearing as it did
in the numerator of her equation
(where, raised to a very high power
indeed, it compensated
for various fault factors
in the denominator), had much
to do with it. But man
does not live on bed alone,
and the import of other variables,
equally complex, does not escape her.

She considers the possible solutions,
real and imaginary,
all neatly arrayed on the foolscap
of her cerebral cortex,
and realizes with dismay that,
despite their abiding love
for one another, she and her husband
had chosen asymptotic pathways
through life's topography:

they would always grow closer
and closer, the chasm between them
getting smaller and smaller forever,
but would never quite coincide.

She smiles, recognizing a good thing
when she sees it. Besides, not being
numbers, they could always

reach across the gap.

## From *A Mathematical Madness*

For months I had been "grinding" Mathematics day and night
When Miss McGirton cast on my affections such a blight:
My mind unhinged now only creaks, and when I tell my woes
I'm forced to lisp in *numbers* what I'd rather say in prose.

Sweet maiden *perpendicular!* She gave a *slanting* sigh
As o'er my kneeling form she cast a calculating eye.
"Ah! well," said I, "you *cipher* me, for if you'll not be mine
From out this pocket next my heart I'll *straight produce a line;*
So ere you are, dear *Polly, gone,* pray heed your lover's vow,
Or he dangles at *right angles* to some *horizontal* bough."

The maid flew in no *frustum*—like your giddy gushing girls—
But standing calm and frigid, shook her strictly *spiral* curls,
And said, "You see we're equal as to station: very well!
*Our paths in life could never meet, because they're parallel.*"

Her voice was so serrated that I fled this maid antique;
Then, approaching her *obliquely, at a tangent* took her cheek!
The kiss was too *elliptical!* She vanished into space!
And a circulating obelisk now marks the fatal place.

# The Proof

I could live like this, waiting on the roof
for the great egret that flies overhead
at just this time, measuring the sun's height
with my fingers to see if the moment's come,
Annie studying the horizon as she describes
the last minutes of a show she watched
in which some mathematician—
she didn't catch the name—labours seven years
to solve a proof he's been enthralled by
since childhood, and though Annie tuned in
too late to know the nature of the problem,
she loves the pure joy with which he looks
into the camera and announces, *I've found it*—
there are tears in his eyes—*I've found it.*

A fishing dragger's heading home
with its trailing veil of gulls, the egret's
late in coming, and it turns out that
after presenting his proof to his colleagues,
after being feted for the discovery of the century,
the mathematician sees one day his solution
is flawed. Even recounting this, Annie's face
is radiant. The poor guy buries himself in his room
another year, a friend visits to tell him it's a gift
to be able to make good mistakes, it's harder
than you'd think to make mistakes that lead
to something new, but our hero's unconsolable.
Finally, his wife says the only thing she wants
for her birthday is the correct proof,
but he can't solve it, even for her.
As Annie speaks, gulls keep swooping
close behind her and then gliding up
over the flat upper roof to drop clams
and crack them open, and yes, our mathematician
at last errs dazzlingly and stumbles upon

his new theorem, and no, we don't know
who this gentle smiling man is or what he's found,
but the evening coming on is filled with such
sweetness that we're up and all but dancing
as we dedicate ourselves to making
a few glorious mistakes and watch
one scruffy gull eye us from the gable
as he guards his food fiercely, fiercely.

James Dickey

## *Math*

*—Lautréamont—*

Numbers who can't ever hear me
I'll say it anyway
All the way from my age-old school. You're still in my heart,
And I can feel you go through there
Like a clean sea-wave. I breathed-in, instinctively,
From the one-two, one-two counts
Of the soft-rocking cradle

As drinking from a universal spring

older than the sun:

Numbers. There is this wave of matched, watched numbers
In my school-soul. Sometimes it is like smoke: I can't get through it.
Sometimes I believe that you've put put in place of my heart
Inhuman logic. Coldness
beyond bearing. And yet . . . because of you
My intelligence has grown far beyond me
from the frozen, radiant center
Of that ravishing clarity you give: give to those
Who most truly love you and can find you: *Listen,* ever-deaf numbers.
Hail! *I* hail you
Arithmetic! Algebra! Geometry!

Triangle gone luminous!

## *Geometry*

I prove a theorem and the house expands:
the windows jerk free to hover near the ceiling,
the ceiling floats away with a sigh.

As the walls clear themselves of everything
but transparency, the scent of carnations
leaves with them. I am out in the open

and above the windows have hinged into butterflies,
sunlight glinting where they've intersected.
They are going to some point true and unproven.

John L. Drost

## I Even Know of a Mathematician

*"I even know of a mathematician who slept with his wife only on prime-numbered days..." Graham said.*

*—Paul Hoffman, The Man Who Loved Only Numbers*

A mathematician was obsessed with things prime.
He thought about them almost all of the time.
Said to his dear wife, "It truly seems right
That we should only make love on a prime-numbered night."
His wife thought for a bit ('cause she was no dummy),
"At the month's start this does seem quite yummy,
For there's two, three, five, seven
A three-night hiatus and then there's eleven.
But of the month's end I start to be wary
Near the twenty-third day of the month February.
For the next prime day after will be March the first
Such sexual continence might cause me to burst!"
He shook his head sadly, "As it's commonly reckoned,
The next prime day would be found on the second."

# *The Mathematician, Counting*

One cannot help loving a mathematician.
Such guts, such pizzaz!
Plump Copernicus, full of round theories;
Euclid, pondering his postulates
and counting, always counting
toward a finite infinity.

All the while, the heart's
little time bomb is busy,
adding and subtracting.
I have seen both the sun
and the moon in the morning
What does this prove?
Some superstition?
Some old astronomer's tale?
No, no . . . only the proof of our spinning—
the geometry of the universe.

It is August.
The meadows are afloat
with Queen Anne's Lace,
each a starry galaxy.
We are an equation devoted to itself,
a myriad of seeds, lives, stars,
bursting their pods,
each capable, countable,
inchoate,
an endless census,
a mathematical fecundity
in which decades are the norm.
And each is devoted to the count,
always the count, the counting
theory holds, even if infinity should fail.

Cathryn Essinger

## *You Are Right*

In your super-logical,
analytical,
bumbling way,
with halting speech
and much digression,
you explain that male
mathematicians are rarely
verbal . . .

> "oh, by the way,
> did I mention that this theory
> is largely unproven, but
> nevertheless, quite probable?"

because of a pre-natal
super-dose of testosterone
to the left side of the brain
which suppresses the right
side of the brain
were you are currently
trying to express
your lack of verbal agility

while at the same time
peeling an orange,
stroking your mustache,
pulling your ear,
and making little
finger-steeples

And I am about to conclude
that you are right.

## Calculus

I tell my students the story of Newton versus Leibniz,
the war of symbols, lasting five generations,
between The Continent and British Isles,
involving deeply hurt sensibilities,
and grievous blows to national pride;
on such weighty issues as publication priority
and working systems of logical notation:
whether the derivative must be denoted by a "prime,"
an apostrophe atop the right hand corner of a function,
evaluated by Newton's fluxions method, $\Delta y / \Delta x$;
or by a formal quotient of differentials $dy/dx$,
intimating future possibilities,
terminology that guides the mind.
The genius of both men lies in grasping simplicity
out of the swirl of ideas guarded by Chaos,
becoming channels, through which her light poured clarity
on the relation binding slope of tangent line
to area of planar region lying below a curve,
The Fundamental Theorem of Calculus,
basis of modern mathematics, claims nothing more.

While Leibniz—suave, debonair, philosopher and politician,
published his proof to jubilant cheers of continental followers,
the Isles seethed unnerved,
they knew of Newton's secret files,
locked in deep secret drawers—
for fear of theft and stranger paranoid delusions,
hiding an earlier version of the same result.
The battle escalated to public accusation,
charges of blatant plagiarism,
excommunication from The Royal Math. Society,
a few blackened eyes,
(no duels);
and raged for long after both men were buried,

splitting Isles from Continent, barring unified progress,
till black bile drained and turbulent spirits becalmed.

Calculus—Latin for small stones,
primitive means of calculation; evolving to abaci;
later to principles of enumeration advanced by widespread use
of the Hindu-Arabic numeral system employed to this day,
as practiced by *algebristas*—barbers and bone setters in Medieval Spain;
before Calculus came the $\sum$ (sigma) notion—
sums of infinite yet countable series;
and culminating in addition of uncountable many dimensionless line segments—
the integral $\int$—snake,
first to thirst for knowledge, at any price.

That abstract concepts, applicable—at start,
merely to the unseen unsensed objects: orbits of distant stars,
could generate intense earthly passions,
is inconceivable today;
when Mathematics is considered a dry discipline,
depleted of life sap, devoid of emotion,
alive only in convoluted brain cells of weird scientific minds.

Sarah Glaz

## *Late Afternoon at the Workshop on Commutative Rings*

*Il Palazzone, Cortona, Italy, June 2006*

*The nature of a faithful content ideal*
*of a Gaussian polynomial over a commutative ring*
a source of excitement for a number of years
*proves to be locally principal*
The first idea relayed from mind to mind
at last reached perfect formulation

Vino rosso  O  Vino rosso
Brindisi  Salute  Chin-chin  To our quest!
Inebriated in Pienza and Montepulciano
Cortona  Firenze  and  San Gimignano
Amarone del Valpocello  Sangiovese di Romagna Calisto
Chianti Classico Riserva  Akronte  Saltapicchio
Silken on the palate like a Puccini aria
on the lips of Maria Callas

# Douglas Goetsch

## Counting

I'd walk close to buildings counting
bricks, run my finger in the grout
till it grew hot and numb. Bricks
in a row, rows on a floor, multiply
floors, buildings, blocks in the city.
I knew there were numbers for everything—
tires piled in mountains at the dump,
cars on the interstate to Maine,
pine needles blanketing the shoulder of the road,
bubbles in my white summer spit.
I dreamed of counting the galaxies
of freckles on Laura MacNally,
touching each one—*she loves me,*
*she loves me not*—right on up her leg,
my pulse beating away at the sea
wall of my skin, my breath
inhaling *odd,* exhaling *even.*

To know certain numbers
would be like standing next to God,
a *counting* God, too busy
to stop for war or famine.
I'd go out under the night sky
to search for Him up there:
God counting, next to Orion
drawing his bow. I'd seen
an orthodox Jew on the subway,
bobbing into the black volume
in his palms, mouthing words
with fury and precision, a single
drop of spittle at the center
of his lip catching the other lip
and stretching like silk thread.
At night I dreamed a constant stream
of numbers shooting past my eyes so fast
all I could do was whisper as they

came. I'd wake up reading the red
flesh of my lids, my tongue
flapping like ticker tape.

I come from a family of counters;
my brother had 41 cavities in 20 teeth
and he told everyone he met;
Grandpa figured his compound
daily interest in the den, at dusk,
the lights turned off, the ice
crackling in his bourbon; my father
hunched over his desk working
overtime for the insurance company,
using numbers to predict
when men were going to die.

When I saw the tenth digit added
to the giant odometer in Times Square
tracking world population, I wondered
what it would take for those wheels
to stop and reverse. What monsoon
or earthquake could fill graves faster
than babies wriggled out of wombs?
Those vast cemeteries in Queens—
white tablets lined up like dominoes
running over hills in perfect rows—
which was higher, the number
of the living or the dead? Was it
true, what a teacher had said:
get everyone in China to stand on a bucket,
jump at exactly the same time
and it'd knock us out of orbit?
You wouldn't need *everyone*,
just enough, the right number,
and if you knew that number
you could point to a skinny
copper-colored kid and say
*You're the one, you can send us flying.*
That's all any child wants: to count.
That's all I wanted to be, the millionth
customer, the billionth burger sold, the one
with the foul ball, waving for TV.

## *My Dance is Mathematics*

*Down, down, down into the darkness of the grave*
*Gently they go, the beautiful, the tender, the kind;*
*Quietly they go, the intelligent, the witty, the brave.*
*I know. But I do not approve. And I am not resigned.*
*—Edna St. Vincent Millay, Dirge without Music*

They called you *der* Noether, as if mathematics
was only for men. In 1964, nearly thirty years
past your death, at last I saw you in a spotlight,
in a World's Fair mural, "Men of Modern Mathematics."

Colleagues praised your brilliance—but after
they had called you fat and plain, rough and loud.
Some mentioned kindness and good humor
though none, in your lifetime, admitted it was you
who led the way to axiomatic algebra.
Direct and courageous, lacking self-concern,
elegant of mind, a poet of logical ideas.

At a party when you were eight years old
you spoke up to solve a hard math puzzle.
Fearless, you set yourself apart.

I followed you. I saw you forced to choose
between mathematics and other romance.
For women only, this exclusive standard.

I heard fathers say, "Dance with Emmy—
just once, early in the evening. Old Max
is my friend; his daughter likes to dance."

If a woman's dance is mathematics,
she dances alone.

Mothers said, "Don't tease. That strange one's heart
is kind. She helps her mother clean the house,
and cannot help her curious mind."

Teachers said, "She's smart but stubborn,
contentious and loud, a theory-builder
not persuaded by our ideas."

Students said, "She's hard to follow, bores me."
A few stood firm and built new algebras
on her exacting formulations.

In spite of Emmy's talents,
always there were reasons
not to give her rank
or permanent employment.
She's a pacifist, a woman.
She's a woman and a Jew.
Her abstract thinking
is female and abstruse.

Today, history books proclaim that Noether
is the greatest mathematician
her sex has produced. They say she was good
for a woman.

## *San Antonio, January, 1993*

A mathematician left the convention
focused on 9, the digit that sits
in the billionth decimal place of pi,
ratio of circumference to width
of the yellow circle that parted the clouds
as she strolled down Commerce Street
to the Rio Rio Café for lunch and a beer.

On fire with jalapeños
she went shopping
for a souvenir.
She bought earrings—
red-red plastic peppers
with green stems.

She said, "Hot peppers
are like mathematics—
with strong flavor
that takes over
what they enter."

## Haipeng Guo

# *When a P-Man Loves an NP-Woman*

Being a happy deterministic man
With a simple polynomial brain
I contented myself with P problems,
And always looked at NP with disdain.

Fell in love with a polynomial woman,
With a non-deterministic wit.
She said she would marry me,
Only if I could show her that P = NP.

I rushed to the library and studied,
Asked *Garey & Johnson* for a hint to the truth.
They said "This is quite a hard question,"
And none had a hint or a clue.

Went to church and prayed to the Almighty,
"Please, Lord, give a lead to the truth,"
"Don't waste your time son," a voice said laughing,
"For on this question I too wasted my youth."

First oracle says you will marry,
Second one tells you you'll split.
Time moves, paths branch, results vary—
Accept the state that finally fits.

"If at last you marry this girl,
And P = NP holds true—such chaos:
e-banking's unsafe, salesmen can travel cheap!—
And mathematicians have nothing to do!

"If I consent to grant you happiness,
A precondition is—no witnesses.
Although you each did nothing wrong,
The punishments will be exponentially long.

"If you really want to marry this woman,
Then randomness might be your only key,
But please stop praying for an answer from me,
For I cannot decide whether P truly equals NP!"

## "...White of Forgetfulness, White of Safety"

My mother was burning in a closet.

Creekwater wrinkling over stones.

Sister Damien, in fifth grade, loved teaching mathematics.
Her full white sleeve, when she wrote on the board,
Swayed like the slow movement of a hunting bird,
Egret in the tidal flats,
Swan paddling in a pond.

Let A equal the distance between *x* and *y*.

The doves in the desert,
Their cinnamon coverts when they flew.

People made arguments. They had reasons for their appetites.
A child could see it wasn't true.

In the picture of the Last Supper on the classroom wall,
All the apostles had beautiful pastel robes,
Each one the color of a flavor of sherbet.

A line is the distance between two points.

A point is indivisible.

Not a statement of fact; a definition.

It took you a second to understand the difference,
And then you loved it, loved reason,
Moving as a swan moves in a mill stream.

I would not have betrayed the Lord
Before the cock crowed thrice,
But I was a child, what could I do
When they came for him?

Ticking heat, the scent of sage,
Of pennyroyal. The structure of every living thing
Was praying for rain.

# A Lady and Her Lingerie

His lady was lovely, with glamour galore.
She led him one day to a lingerie store.
A dozen of these and two dozen of those
Transparently feminine silk so-and-so's
Would cost him, they said, six bucks less, if you please,
Than a dozen of those and two dozen of these.
Such extravagance, though, was way out of his reach.
He'd only six bucks, which just bought one of each.
Ignoring the frowns and "tut-tuts" of the grundies,
Just figure the price of each item of undies.

Bob Kurosaka

## *There Was a Young Maiden*

There was a young maiden named Lizt
Whose mouth had a funny half-twist.
  She'd turned both her lips
  Into Möbius strips . . .
'Til she's kissed you, you haven't been kissed!

# There's a Delta for Every Epsilon

*Lyrics for a Calypso song*

There's a delta for every epsilon,
It's a fact that you can always count upon.
There's a delta for every epsilon
    And now and again,
    There's also an *N*.

But one condition I must give:
The epsilon must be positive
A lonely life all the others live,
    In no theorem
    A delta for them.

How sad, how cruel, how tragic,
How pitiful, and other adjec-
Tives that I might mention.
The matter merits our attention.
If an epsilon is a hero,
Just because it is greater than zero,
It must be mighty discouragin'
To lie to the left of the origin.

This rank discrimination is not for us,
We must fight for an enlightened calculus,
Where epsilons all, both minus and plus,
    Have deltas
    To call their own.

## *Sacrifice and Bliss*

$$\text{Love} = \lim_{\text{Ego} \to 0} \frac{1}{\text{Ego}}$$

## A Lecture on Thomson's Galvanometer

*Delivered to a single pupil in an alcove with drawn curtains*

The lamp-light falls on blackened walls,
And streams through narrow perforations;
The long beam trails o'er pasteboard scales,
With slow-decaying oscillations.
Flow, current! flow! set the quick light-spot flying!
Flow current! Answer, light-spot! flashing, quivering, dying.

O look! how queer! how thin and clear,
And thinner, clearer, sharper growing,
The gliding fire, with central wire
The fine degrees distinctly showing.
Swing, magnet! swing! advancing and receding;
Swing, magnet! answer, dearest, what's your final reading?

O love! you fail to read the scale
Correct to tenths of a division;
To mirror heaven those eyes were given,
And not for methods of precision.
Break, contact! break! set the free light-spot flying!
Break, contact! rest thee, magnet! swinging, creeping, dying.

## A *Lady of 80*

A lady of 80 named Gertie
Had a boyfriend of 60 named Bertie.
    She told him emphatically
    That viewed mathematically
By modulo 50 she's 30.

# Useless

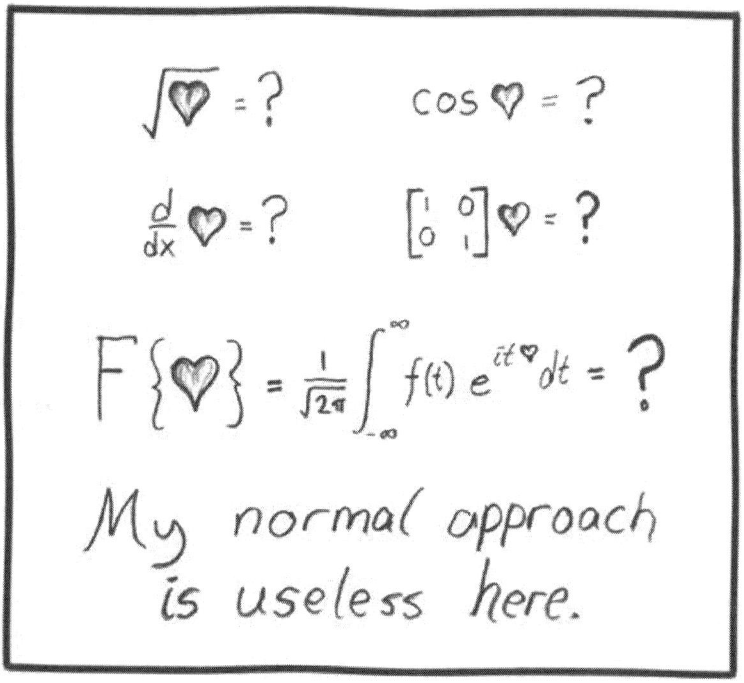

$\sqrt{♥} = ?$
$\cos ♥ = ?$
$\frac{d}{dx} ♥ = ?$
$\begin{bmatrix} 1 & 0 \\ 0 & 1 \end{bmatrix} ♥ = ?$
$F\{♥\} = \frac{1}{\sqrt{2\pi}} \int_{-\infty}^{\infty} f(t)\, e^{it♥} dt = ?$
My normal approach is useless here.

# Figures of Thought

To lay the logarithmic spiral on
Sea-shell and leaf alike, and see it fit,
To watch the same idea work itself out
In the fighter pilot's steepening, tightening turn
Onto his target, setting up the kill,
And in the flight of certain wall-eyed bugs
Who cannot see to fly straight into death
But have to cast their sidelong glance at it
And come but cranking to the candle's flame—

How secret that is, and how privileged
One feels to find the same necessity
Ciphered in forms diverse and otherwise
Without kinship—that is the beautiful
In Nature as in art, not obvious,
Not inaccessible, but just between.

It may diminish some our dry delight
To wonder if everything we are and do
Lies subject to some little law like that;
Hidden in nature, but not deeply so.

## Ode to Numbers

Oh, the thirst to know
how many!
The hunger
to know
how many
stars in the sky!

We spent
our childhood counting
stones and plants, fingers and
toes, grains of sand, and teeth,
our youth we passed counting
petals and comets' tails.
We counted
colors, years,
lives, and kisses;
in the country,
oxen; by the sea,
the waves. Ships
became proliferating ciphers.
Numbers multiplied.
The cities
were thousands, millions,
wheat hundreds
of units that held
within them smaller numbers,
smaller than a single grain.
Time became a number.
Light was numbered
and no matter how it raced with sound
its velocity was 37.
Numbers surrounded us.
When we closed the door
at night, exhausted,
an 800 slipped

beneath the door
and crept with us into bed,
and in our dreams
4000s and 77s
pounded at our foreheads
with hammers and tongs.
5s
added to 5s
until they sank into the sea or madness,
until the sun greeted us with its zero
and we went running
to the office,
to the workshop,
to the factory,
to begin again the infinite
1 of each new day.

We had time, as men,
for our thirst slowly
to be sated,
the ancestral desire
to give things a number,
to add them up,
to reduce them
to powder,
wastelands of numbers.
We
papered the world
with numbers and names,
but
things survived,
they fled
from numbers,
went mad in their quantities,
evaporated,
leaving
an odor or a memory,
leaving the numbers empty.

That's why
for you
I want *things*.
Let numbers
go to jail,
let them march
in perfect columns
procreating
until they give the sum
total of infinity.
For you I want only
for the numbers
along the road
to protect you
and for you to protect them.
May the weekly figure of your salary
expand until it spans your chest.
And from the 2 of you, embraced,
your body and that of your beloved,
may pairs of children's eyes be born
that will count again
the ancient stars
and countless
heads of grain
that will cover a transformed earth.

Translated from the Spanish by Margaret Sayers Peden

Katharine O'Brien

## *Undefined Terms*

A point is a point, a line is a line,
  a rose is a rose is a rose.
We thus undefine in the manner of Stein
  some terms in unrhyme and unprose.

On these as foundation we lay definitions,
  the girders for walls and a roof.
We assume some conditions to fit requisitions
  and build us a logical proof.

When Hilbert found flaws in the structural scene,
  he repaired the unrigor discreetly.
We know what we mean when we speak of "between,"
  which Euclid unnoticed completely.

As we trace mathematics up from its roots
  to the reach of its vast propositions,
the beautiful fruits of our lively pursuits
  all stem from our undefinitions.

## Katharine O'Brien

# *Valentine*

You disintegrate my differential,
    you dislocate my focus.
My pulse goes up like an exponential
    whenever you cross my locus.
Without you, sets are null and void—
    so won't you be my cardioid?

## Bill Parry

# *Alexander's Horned Sphere*

The sea was the first to propose it,
anticipating Alexander's thought.
The whited branch which clawed the ocean floor,
the echoing shell, the time-worn stone,
anemone and jell of sea-flower;
had he seen all this before?
Did he find his inspiration
along this sepulchered beach
or in some anatomy lesson?

Strands of bladderwrack,
distend to dentritic kelp. He saw
this could go on forever,
each increment a crab's claw.

So friend, hand in hand, we must keep close.
For should you stray,
perhaps along that promontory,
there's no going back.
Our lonely paths shall intersect
only by leap of synapse.

---

Note from the editors: Alexander's horned sphere image is taken from: J. W. Alexander, *An Example of a Simply Connected Surface Bounding a Region which is not Simply Connected*, Proceedings of the National Academy of Sciences 10 (1924), 8–10.

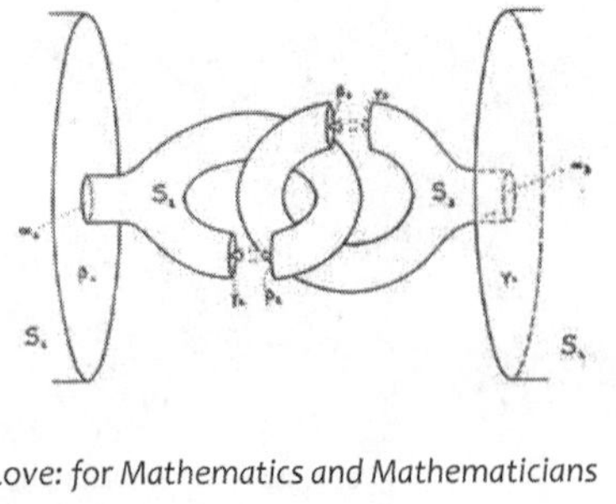

# Argument

As he cleaned the board,
chalk-dust rose like parched mist.
A dry profession, he mused as morosely
they shuffled settling tier upon tier.

Now, almost half-way through the course,
(coughs, yawns and automatic writing)
the theorem is ready.

Moving to the crucial point,
the sly unconventional twist,
a quiver springs his voice and breast;

soon the gambit will appear
opposed to what's expected.
The ploy will snip one strand
the entire skein sloughing to the ground.

His head turns sympathetically
from board to class.
They copy copiously.
But two, perhaps three pause and frown,

wonder will this go through,
questioning this entanglement
—yet they nod encouragement.
Then the final crux; the ropes relax and fall.

His reward: two smile, maybe three,
and one is visibly moved.
Q.E.D., the theorem is proved.

This was his sole intent.
Leaving the symbols on the board
he departs with a swagger of achievement.

## Raymond Queneau

### *Swigns*

when One made love to Zero
the spheres embraced the tori
the first numbers came forward
stretching out their hands towards the fresh sycamores
and the continued fractions (fatally mauled
by a torrent of mute decimals) went to bed

when B made love to A
the paragraphs fell into a wild passion
the commas came forward
stretching out their necks over the iron bridges
and the alphabet (fatally mauled)
fainted in the arms of a mute question

Translated from the French by Joyce S. Weiner and Jet Foncannon

## The Mathematician in Love

I

A mathematician fell madly in love
With a lady, young, handsome, and charming:
By angles and ratios harmonic he strove
Her curves and proportions all faultless to prove.
As he scrawled hieroglyphics alarming.

II

He measured with care, from the ends of a base,
The arcs which her features subtended:
Then he framed transcendental equations, to trace
The flowing outlines of her figure and face,
And thought the result very splendid.

III

He studied (since music has charms for the fair)
The theory of fiddles and whistles,—
Then composed, by acoustic equations, an air,
Which, when 'twas performed, made the lady's long hair
Stand on end, like a porcupine's bristles.

IV

The lady loved dancing:—he therefore applied,
To the polka and waltz, an equation;
But when to rotate on his axis he tried,
His centre of gravity swayed to one side,
And he fell, by the earth's gravitation.

V

No doubts of the fate of his suit made him pause,
For he proved, to his own satisfaction,

That the fair one returned his affection;—"because,
"As every one knows, by mechanical laws,
"Re-action is equal to action."

VI

"Let *x* denote beauty,—*y*, manners well-bred,—
"*z*, Fortune,—(this last is essential),—
"Let *L* stand for love"—our philosopher said,—
"Then *L* is a function of *x*, *y*, and *z*,
"Of the kind which is known as potential."

VII

"Now integrate *L* with respect to *dt*,
"(*t* Standing for time and persuasion);
"Then, between proper limits, 'tis easy to see,
"The definite integral *Marriage* must be:—
"(A very concise demonstration)."

VIII

Said he—"If the wandering course of the moon
"By Algebra can be predicted,
"The female affections must yield to it soon"—
—But the lady ran off with a dashing dragoon,
And left him amazed and afflicted.

## *We Sat, So Patient*

We sat, so patient, in that third
  grade
class, learning the numbers of days,
  weeks, months,
repeating the numbers as they flashed
  in the air, forming
the curved 3, the angular 4, the easy 1,
adding them up, subtracting, multiplying
  and dividing
as though we owned them, and we did,
  then,
counting the rain drops that wriggled down
  the gray window,
counting the hearts and cars on our desks,
  our crayons,
Ann Harden and Richie Freeman making 2,
10 on each side for the spelling bee,
  counting silent
seconds when Sister Ann Zita said 5
  of us
would not reach 20, showed the chart
where children dropped off into 0,
  the blue zone of No Return
that Jimmy Legasse whispered, making us laugh.
Looking around, I thought Al Aubon, Jackie Foster,
  Dorothy Blake
who already coughed blood on her gold glasses
  when she spoke,
the thin girl just come over from Germany,
and Ray Martineau who had no lunch, the
  zigzagged
white lines of lice forming mazes on his
  crewcut head. And
the good Sister herself, number 6, at least
  40 years older than us,

her rosary beads clicking as she walked
down the aisle like the Angel of Death,
  black
wings spread, brushing our faces, our arms,
  wafting blackness
into our eyes, our lungs, our hearts,
reminding us that God was watching and could tell
  who knew 9 times 9, 144 divided by 12,
telling us it was God's will that we die,
Jimmy Gleason pulling up his white sock
on a leg he would not have 10 years later,
Barbara Ryan raising her hand with another
  correct answer,
the same hand so whitish-blue as she lay
  in her eighth-grade coffin,
Jimmy Amyot and Donald Wilcox quietly passing
  drawings of naked women
back and forth, the car they would die in
  revving
unheard in that classroom where we yelled
  out 360, 225, 32, 0, 10,
waiting for the split-second flash of red,
  yellow and blue cards
beneath the slow, steadily clicking clock.

## *Number Man*

*for the ghost of Johann Sebastian Bach*

He was born to wonder about numbers.

He balanced fives against tens
and made them sleep together
and love each other.

He took sixes and sevens
and set them wrangling and fighting
over raw bones.

He woke up twos and fours
out of baby sleep
and touched them back to sleep.

He managed eights and nines,
gave them prophet beards,
marched them into mists and mountains.

He added all the numbers he knew,
multiplied them by new-found numbers
and called it a prayer of Numbers.

For each of a million cipher silences
he dug up a mate number
for a candle light in the dark.

He knew love numbers, luck numbers,
how the sea and the stars
are made and held by numbers.

He died from the wonder of numbering.
He said good-by as if good-by is a number.

## From *Borderline—A Fractal Poem*

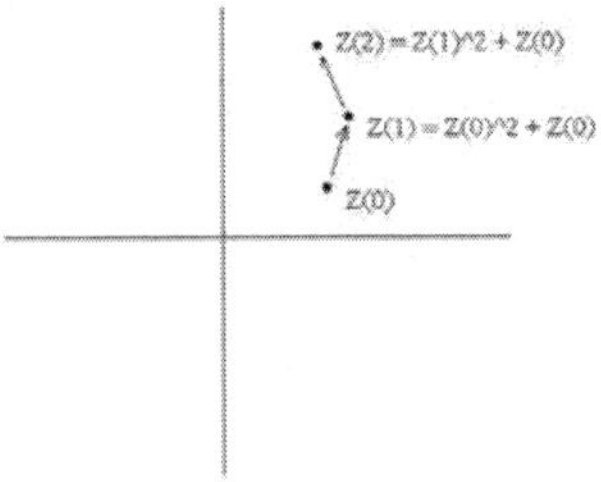


The Mandelbrot Algorithm…

Z-naught, a point is squared and when
It's added to Z-naught again,
Z-one, the sum, in turn is squared
And with Z-naught again it's paired.
For generating where Z is,
One uses this recursive biz.

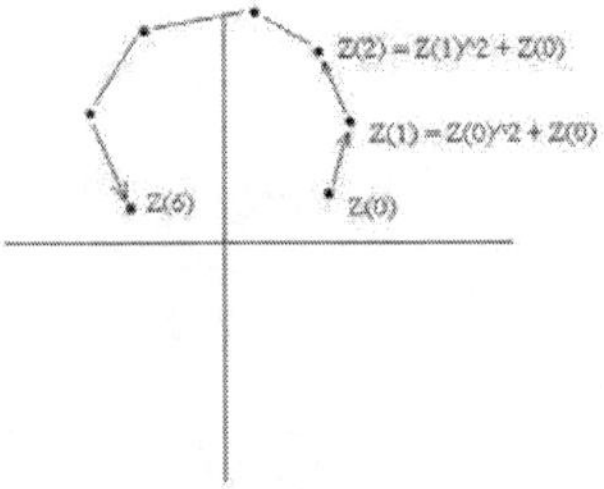


… generates a series of points

Some points escape all gravity
And navigate infinity
While some born closer to the core
Contract and stay there evermore
And some, a very special kind
Hang out along the borderline.

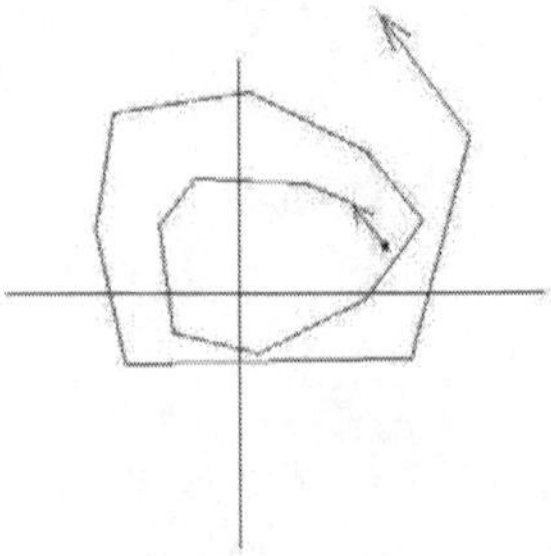

Unstable— heading to infinity

This little game of vector boat
Gets its name from Mandelbrot

Who's set upon the complex plane
Reflects a most complex terrain,
A captivating, endless scene
So ordered, yet so serpentine.

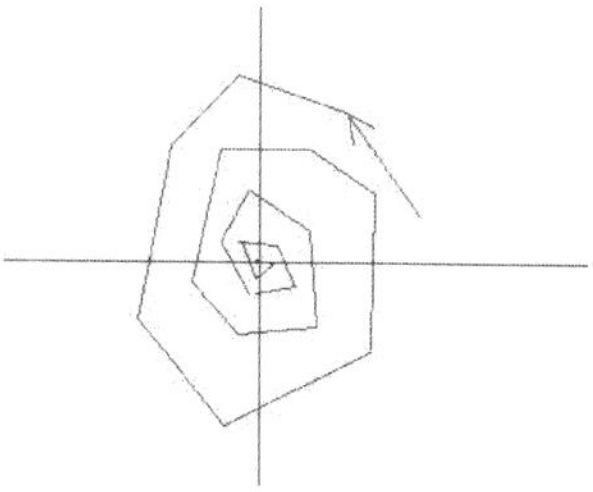

So I begin to walk its trails
And come to know where knowing fails
And logic, coping with the joke
To no avail, finds cosmic smoke
So this my lab-book diary
Comes to be writ in poetry.
. . .

Stable— heading to zero

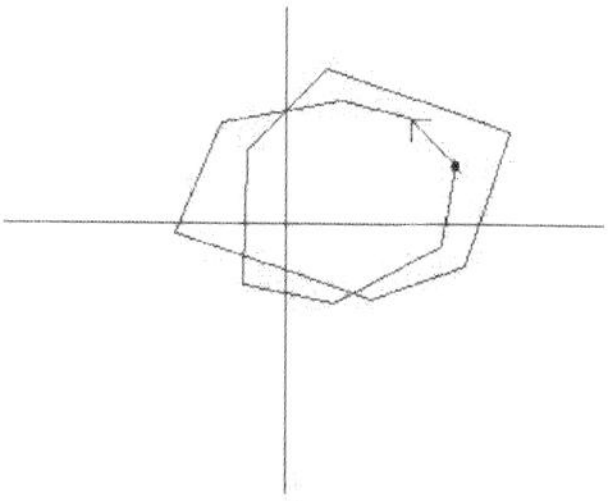

Borderline Stable— staying in a range

The Mandelbrot Set is the set of these Z(0)'s.

# Frederick Soddy

## The Kiss Precise

For pairs of lips to kiss maybe
Involves no trigonometry.
'Tis not so when four circles kiss
Each one the other three.
To bring this off the four must be
As three in one or one in three.
If one in three, beyond a doubt
Each gets three kisses from without.
If three in one, then is that one
Thrice kissed internally.

Four circles to the kissing come.
The smaller are the benter.
The bend is just the inverse of
The distance from the center.
Though their intrigue left Euclid dumb
There's now no need for rule of thumb.
Since zero bend's a dead straight line
And concave bends have minus sign,
*The sum of the squares of all four bends*
*Is half the square of their sum.*

To spy out spherical affairs
An oscular surveyor
Might find the task laborious,
The sphere is much the gayer,
And now besides the pair of pairs
A fifth sphere in the kissing shares.
Yet, signs and zero as before,
For each to kiss the other four
*The square of the sum of all five bends*
*Is thrice the sum of their squares.*

---

Note from the editors: A sequel, by Thorold Gosset, and a note regarding both poems appear on the following page.

## The Kiss Precise (Generalized)

And let us not confine our cares
To simple circles, planes and spheres,
But rise to hyper flats and bends
Where kissing multiple appears,
In n-ic space the kissing pairs
Are hyperspheres, and Truth declares—
As n + 2 such osculate
Each with an n + 1 fold mate
*The square of the sum of all the bends*
*Is n times the sum of their squares.*

Note from the editors: Frederick Soddy (1877–1956), Nobel prize winning British chemist, rediscovered the Descartes Circle Theorem, which involves the radii of four mutually tangent circles. The final stanza of Soddy's poem announces Soddy's discovery of an analogous formula for spheres. After Soddy's verses appeared, Thorold Gosset wrote additional lines to describe the more general case for tangency, or "kissing," of n + 2 hyperspheres in n dimensions.

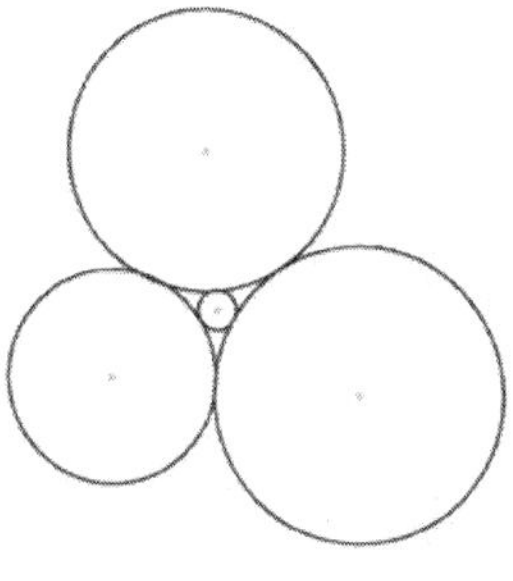

## *Let Us Now Praise Prime Numbers*

Let us now praise prime numbers
With our fathers who begat us:
The power, the peculiar glory of prime numbers
Is that nothing begat them,
No ancestors, no factors,
Adams among the multiplied generations.

None can foretell their coming.
Among the ordinal numbers
They do not reserve their seats, arrive unexpected.
Along the lines of cardinals
They rise like surprising pontiffs,
Each absolute, inscrutable, self-elected.

In the beginning where chaos
Ends and zero resolves,
They crowd the foreground prodigal as forest,
But middle distance thins them,
Far distance to infinity
Yields them rare as unreturning comets.

O prime improbable numbers,
Long may formula-hunters
Steam in abstraction, waste to skeleton patience:
Stay non-conformist, nuisance,
Phenomena irreducible
To system, sequence, pattern or explanation.

# *Striving All My Life*

Maxwell said: There is no more powerful way
to introduce knowledge to the mind than ... as many different
ways as we can, wrenching the mind

away
from the symbols to the objects and from the objects
back to the symbols.

Maxwell said: I have been striving all my life to be free
of the yoke of Cartesian co-ordinates. I found
such an instrument in

quaternions. Do I need *quaternions*
to talk about light?
Alas,

the square of quaternions
is negative. But Gibbs's vectors, uncouth
seemingly, work

well, in *any* dimension, with a very
great capability for
interpreting space relations.

Rukeyser said: Critical minds
that approach the world with love
have but one possible

defense—to build a system.
Rayleigh said, I protest
the compression.

Gibbs: I myself concluded
that the paper was
too long.

Michael Stueben

# Valentine

You are the fairest of your sex,
Let me be your hero,
I love you as one over x,
As x approaches zero.

## Unknown Author

### *Puzzle*

I am obliged to plant a grove,
To please the pretty girl I love.
This curious grove I must compose
Of nineteen trees in nine straight rows;
And in each row, five trees must place,
Or I may never see her face.
Now, readers brave, I'm in no jest.
Pray lend your aid and do your best.

Alissa Valles

## *Mathematician*

At twenty, I got drunk on Friday nights
at the conservative students' association.
I did analysis on beer mats with a felt pen,
sometimes continuing onto the table.
Once I won a bet with my professor; I
covered two tables and the seat of a chair.
I learned that a voice without undertones
thrown from the back of the mouth draws
a person's attention across several rooms.

At thirty, an associate professor, I lived
with a blond girl in a sunny apartment
by a park and spent nights tracing the fits
of passion and rage in Beethoven. I tried
to explain to her my notation system
for the *Pathétique*, which puts sequences
of notes, trills, and dynamic shifts into
equations of exquisite formal elegance.
Last year, no longer a girl, she left.

I lay on the bed not crying or laughing.
I bought oranges and let them rot in the nets.
The melody that follows the tolling bells
in Debussy's *Cathédrale engloutie* is
the most beautiful thing I've ever heard;
a sea rises again and again to drown it out.
I'm taking singing lessons to learn how
to project from the very front of the palate.
I want to express every shade of feeling.

I've begun to speak with a softer voice,
gentler, more alone, which is what one
should be. I bought a TV and spend nights
watching the war. I want to know how
to find an oasis of tenderness in a desert

of egotism. I see only the familiar mirages
of satisfaction, encampments of convention.
I'm now working on a symbolic language
which will enable me to elucidate all this.

# *The Song of 10*

From the Romans' *decem* our decibels and decimal system, O tenfold
the sorrows of Israel, Decameron tales meant to be told over ten nights

in December, solstice month frozen in moondrifts of snow. Our fingers
and toes. Kingly ten-pointed stags reigning over Europe's greenwoods,

for miners a measure in tons of coal or type of tallow candle weighted
ten per pound, the legion poor mending by its light. What else is there

to say? Higher than nine. A number whose power is mighty to multiply,
comprising one and nil, wand and egg, gold spindle and heavenly wheel

of goddess Fate who turns time and tides; what our parents say summer
evenings, hearing our voices dart and flicker in neighboring yards before

we dance from them into darkness and love's rule ends—*I'll count to ten.*

# Aigeltinger

In the bare trees old husks make new designs
Love moves the crows before the dawn
The cherry-sun ushers in the new phase

The radiant mind
addressed by tufts of flocking pear blossoms
proposes new profundities to the soul

Deftness stirs in the cells
of Aigeltinger's brain which flares
like ribbons round an electric fan

This is impressive, he will soon proclaim
God!

And round and round, the winds
and underfoot, the grass
the rose-cane leaves and blackberries
and Jim will read the encyclopedia to his
new bride—gradually

Aigeltinger you have stuck in my conk
illuminating, for nearly half a century I
could never beat you at your specialty

Nothing has ever beaten a mathematician
but yeast

The cloudless sky takes the sun in its periphery
and slides its disc across the blue

They say I'm not profound

But where is profundity, Aigeltinger
mathematical genius
dragged drunk from some cheap bar to serve
their petty purposes?

Aigeltinger, you were profound

# Clarence R. Wylie, Jr.

## *Paradox*

Not truth, nor certainty. These I forswore
In my novitiate, as young men called
To holy orders must abjure the world.
"*If* . . . , *then* . . . ," this only I assert;
And my successes are but pretty chains
Linking twin doubts, for it is vain to ask
If what I postulate be justified,
Or what I prove possess the stamp of fact.

Yet bridges stand, and men no longer crawl
In two dimensions. And such triumphs stem
In no small measure from the power this game,
Played with the thrice-attenuated shades
Of things, has over their originals.
How frail the wand, but how profound the spell!

# *Contributors' Notes*

*Diane Ackerman* is author of eleven books of nonfiction, including the best seller *A Natural History of The Senses* and seven volumes of poetry. Her work received the John Burroughs Nature Award, the Lavan Poetry Prize, and other honors. She holds a PhD in English from Cornell and has taught at various universities, including Cornell and Columbia.

*Rafael Alberti* (1902–1999) was a Spanish poet, playwright, and painter. He wrote of painting and poetry as complementary arts. He is author of 24 volumes of poetry, including his best known, *Concerning the Angels*, written in late 1920s at the height of Madrid's literary renaissance. Alberti was awarded the *Premio Cervantes*, the Spanish literary world's highest honor, in 1983.

*Sandra Alcosser* is National Endowment for the Arts' first Conservation Poet and Montana's first poet laureate. Her seven books of poetry have won numerous honors including the National Poetry Series, the Associated Writing Programs Award Series, and the Academy of American Poets James Laughlin Award. About mathematics she says, "Resource analysis is a form of love."

*Manuel Alegre* is a Portuguese writer and political figure. He studied law at Coimbra University; his political views led to periods of imprisonment and exile. Since returning to Portugal in 1974 he has been active in the Socialist Party and was a candidate for presidency in 2006. His poem, "Uma flor de verde pinho," set to music, represented Portugal in the Eurovision Song Contest (1976).

*Karren LaLonde Alenier*, a veteran computer programmer/ analyst/ specialist, wrote *Looking for Divine Transportation*, winner of Towson University's Prize for Literature. In 2005, *Gertrude Stein Invents a Jump Early On*, her opera with William Banfield, premiered in New York. In 2007, *The Steiny Road to Operadom: The Making of American Operas*, was published. Visit: alenier.blogspot.com.

*Dante Alighieri* (1265–1321) was the leading European poet of the late Middle Ages. His most famous work, *The Divine Comedy*, is a literary masterpiece and a synthesis of his diverse philosophical views. Dante was born to a noble Florentine family. Involvement in politics

led to exile from Florence and a nomadic existence for the second half of his life.

*Yehuda Amichai* (1924–2000) was born in Germany and immigrated with his family to Palestine in 1936. He made a living as a school teacher. He is Israel's most popular poet as well as an international literary figure. In 1982 he received the Israel Prize for Poetry, his country's highest honor. His work has been translated into more than thirty-five languages.

*Carl Andre* is a well-known sculptor and was a leading member of the Minimalist movement in the mid-sixties. In addition to making sculptures he also creates poems to be exhibited as art objects. His sculptures are in the permanent collections of a number of major museums. He lives and works in New York City.

*Stelian Apostolescu* was for many years a professor of mathematics and physics at the University of Bucharest. His PhD thesis is in nuclear physics: *The Upper Limit of the Electric Dipole of the Neutron*. A translator into English for several Romanian poets, Apostolescu admires in Lucian Blaga the depth of his scientific and philosophical thinking.

W. H. *Auden* (1907–1973) was born in England and immigrated to the US in 1939. One of the most respected poets of the twentieth century, he won the Pulitzer Prize in 1947 for *The Age of Anxiety*. Auden's familiarity with mathematics is illustrated in a number of his poems, including "The Hidden Law," "Numbers and Faces," and "Precious Five."

*Amy Quan Barry* is the author of two books of poetry, both published with University of Pittsburgh Press, *Asylum* (2001), winner of the Agnes Starrett Poetry Prize, and *Controvertibles* (2004). She has been a Wallace Stegner Fellow at Stanford University and is currently Assistant Professor of English at the University of Wisconsin.

*Judith Baumel* is the author of two books of poetry: *The Weight of Numbers*, winner of the Walt Whitman Award, and *Now*. She is Director of Creative Writing at Adelphi University. A graduate of The Bronx High School of Science, she studied chemistry and mathematics at Harvard University before she discovered her true love.

*David Bellos*, born in England, teaches French and comparative literature at Princeton University. He has written widely on nineteenth-century French literature and published several biographies and numerous translations. He holds the rank of Chevalier in the Ordre des Palmes Académiques and received a number of major awards for his work.

*Jakob Bernoulli* (1654–1705), sometimes called Jacques, the first of the Bernoulli family of Swiss citizens to become a mathematician, is known for his contributions to calculus and probability. His pioneering work *Ars Conjectandi,* published posthumously in 1713, contains, in addition to many fine mathematical ideas, philosophical musings and snippets of poetry.

*Bhaskaracharya* (1114–1185) is one of the best known mathematicians of ancient India. His famous work *Siddhanta Siroman* is divided into four parts: *Lilavati* (arithmetic), *Bijaganita* (algebra), *Goladhyaya* (celestial globe), and *Grahaganita* (mathematics of the planets). *Lilavati* means "beautiful"—it is the nickname of his daughter, for whom the work was written.

*Linda Bierds* has published seven books of poetry, her most recent *First Hand* (Penguin, 2005). *Flight: New and Selected Poems* is forthcoming in 2008. Among her many awards are the PEN West Poetry Prize, a Guggenheim fellowship, and a MacArthur Foundation grant. She is Professor of English at the University of Washington.

*Lucian Blaga* (1895–1961) was a Romanian poet and philosopher. In 1949 he was dismissed by the communist regime from his position as professor at the University of Cluj. In 1956, the communist government sent emissaries to protest against his nomination for a Nobel Prize in literature. Out of favor, he published only translations until 1960.

*Ariel Bloch* is Professor Emeritus of Near Eastern Studies at the University of California, Berkeley. His books and articles deal with classical Arabic, Arabic dialectology, biblical and modern Hebrew, Ugaritic, Akkadian, and Aramaic. He is cotranslator of *The Song of Songs.*

*Chana Bloch* is the author of three books of poems: *The Secrets of the Tribe, The Past Keeps Changing,* and *Mrs. Dumpty.* She is cotranslator of *The Song of Songs,* as well as of several books by Israeli poets Yehuda Amichai and Dahlia Ravikovitch. Bloch is Professor Emerita of English Literature at Mills College.

*Ray Bobo* is a retired Georgetown University math professor. His diverse poems have appeared in *The College Mathematics Journal* and *The American Mathematical Monthly.* Ray prefers to find poetry in mathematics rather than mathematics in poetry. Most of his poems are purely mathematical. In this regard, "Epsilon" is not typical of his poetic works. Contact: raybobo@bellatlantic.net.

*David Brooks* is an Australian writer who is fascinated by paradox. His stories include "The Mathematics of Desire" and "The Geometrist" (about parallel lines meeting). His recent poetry collection, *Walking to Point Clear*, was short-listed for major Australian awards. Professor of English at the University of Sydney, he describes his current locations numerically: 033S52/151E13 and 045N31/013E36.

*Elizabeth Barrett Browning* (1806–1861), the most celebrated woman poet of the nineteenth century, married Robert Browning, six years her junior, at the age of 40. The Brownings left England and settled in Florence, where they wrote poetry, had a child, and led a happy life. Her best known work is *Sonnets from the Portuguese*, which chronicles her famous love affair.

*Ann Calandro* is a writer, medical editor, and mixed media artist. She has had fiction and poetry published in various magazines. Her art has been exhibited at numerous venues, including www.jerseyarts.com (April 2007 featured artist). Her mathematical poem was written while she was a graduate student in English at Washington University, St. Louis.

*Enriqueta Carrington* is a Mexican-English writer and mathematician. Her translations and poetry in Spanish and English have appeared in various publications and have been nominated for the Pushcart Prize. She holds a PhD in mathematics from Rutgers University, where she now teaches. Her mathematical work is on Lie groups and representation theory.

*Lewis Carroll* (1832–1898) is the pen name of British mathematician, photographer, and writer Charles Lutwidge Dodgson. Dodgson lectured on mathematics at Oxford University and wrote mathematics books in his own name. He is best known as the author of *Alice in Wonderland* and other children's stories, puzzles, and witty verses.

*Gaius Valerius Catullus* (84–54 BC) is believed to have been born in Verona but lived most of his life in Rome. Catullus fell in love with "Lesbia," believed to be Claudia Metelli, for whom he wrote love poems. Catullus' poetry influenced many of the later celebrated Roman poets and others far removed in time and geography.

*Siv Cedering* is a Swedish-born American writer, poet, translator, composer, painter, and sculptor. She is the author of eighteen books of poetry and fiction in Swedish and English. Her most recent volume is *Letters from an Observatory: New and Selected Poems, 1973–1998* (1998). Her work has received a number of major awards.

*Robin Chapman*, Professor Emerita of Communicative Disorders at the University of Wisconsin, is author of ten collections of poetry, including *Images of a Complex World: The Art and Poetry of Chaos* and, most recently, *The Dreamer Who Counted the Dead* and *Smoke and Strong Whiskey*. She coedited the poetry anthology *On Retirement: 75 Poems*.

*Edward E. Chipman* (1901–2004) was a Baptist minister. A graduate of Colgate University and the Colgate Rochester Divinity School, he taught philosophy and religion at the University of Chicago before entering the ministry. He later developed new church congregations in the US for the American Baptist Home Mission Societies.

*Inger Christensen* is one of Denmark's most honored poets. Her work includes six volumes of poetry, books of fiction, children's books, plays, radio pieces, and numerous essays. She was appointed to the Danish Academy and has received, among other honors, the Grand Prix des Biennales Internationales de Poesie.

*John Ciardi* (1916–1986) was an American writer of Italian origins and author of more than forty volumes of poetry, an acclaimed translation of Dante's *Divine Comedy*, and a textbook for poetry courses. He taught at Harvard and Rutgers Universities, hosted a network television program, and reported for National Public Radio.

*Hanns Cibulka* (1920–2004) was a German poet and diarist. He served in the German armed forces during WWII. After the war he studied library sciences and worked until his retirement as a librarian in Gotha, East Germany. He has published several poetry collections. A selection of his poetry has been translated into English by Ewald Osers (1985).

*Dan Clark*, a former athlete, is CEO of a multimillion-dollar corporation as well as a public speaker, songwriter and recording artist, author, and inspirational personality. He has spoken to numerous audiences in US and abroad, appeared on television, and contributed stories and poems to the *Chicken Soup for the Soul* series and other publications.

*Marion Deutsche Cohen* is a writer and mathematician. She holds a PhD in mathematics from Wesleyan University and teaches at several universities in Philadelphia area. Cohen is the author of several books on spousal illness. Her poetry has appeared in many periodicals and recently in her collection, *Crossing the Equal Sign* (2006).

*Mary Cornish* holds an MFA (2001) in creative writing from Sarah Lawrence College and has received a Stegner Fellowship in Poetry from Stanford University. Her book, *Red Studio* (Oberlin Press, 2007), is the winner of the Field Poetry Prize. She currently teaches at Fairhaven College of Western Washington University.

*Jonathan Coulton* is a singer and songwriter who delivers well-crafted, humorous, geeky folk-pop songs. He studied music at Yale University and is currently Contributing Troubadour for *Popular Science* magazine and musical director for John Hodgman's Little Gray Book Lectures. For more, visit: www.sonicbids.com/jonathancoulton.

*Jennifer Crow* fell in love with poetry at the age of eight and never recovered. Her work has appeared in many print and online venues, and several of her poems received honorable mentions in past editions of *The Year's Best Fantasy and Horror* anthologies. She lives near Buffalo with her family and keeps her distance from mathematics.

J. V. *Cunningham* (1911–1985) drifted through various jobs till he was invited by critic and poet Yvor Winters to study at Stanford University, where he received a PhD in 1945. He briefly taught mathematics to Air Force pilots, during WWII, and taught English at Brandeis University till his retirement. He is author of six books of poetry.

*Keith Allen Daniels* (1956–2001) worked as a materials engineer by day and poet and publisher by night. His poetry awards include a Fallot Literary Award for *What Rough Book* (1992). His collection *Satan is a Mathematician* (1998) was a finalist for the Writer's Digest National Book Award for Poetry.

*Jon Davis* has published five collections of poetry, including *Scrimmage of Appetite*, a winner of the Lannan Literary Award in Poetry. He has recently completed three new volumes and maintains a satirical blog: voydofcourse.blog.com. He says: "My mathematics endeavors are confined to the days immediately proceeding each April 15th."

*Jean de Sponde* (1557–1595) was a French poet of Protestant Basque origin. His love sonnets date from the period in his youth when he wrote and translated secular poetry. At age 25 he became a devout Protestant, and his poetry took on a religious orientation. Persecuted for his religious beliefs, he converted to Catholicism. He died in poverty in Bordeaux.

*Theodore Deppe* is the author of three books of poems, including *Cape Clear: New and Selected Poems* (2002). He directs the Stonecoast in

Ireland MFA program and currently lives in Donegal. His poetry has received a number of awards, including fellowships from the National Endowment of the Arts and a Pushcart Prize.

*Kathryn DeZur* is Associate Professor of English at the State University of New York Delhi College of Technology. Her poems have appeared in *Literary Mama, The Teacher's Voice,* and *Women's Studies: An Interdisciplinary Journal.* Her father is a mathematician, and she inherited great respect for the field from him.

*James Dickey* (1923–1997) was a US army pilot in WWI and Korea and between the wars graduated from Vanderbilt University with degrees in English, philosophy, and astronomy. He received a number of honors for his work, including election as US Poet Laureate, but his popularity truly exploded after the filming of his novel *Deliverance.*

*Emily Dickinson* (1830–1886) lived all her life in her parents' house in Amherst, Massachusetts. Today she is acknowledged as a major poetical voice of her generation, but during her lifetime she succeeded in publishing only 11 of her poems. Dickinson's unedited, entire body of work did not appear till 1955.

*John Donne* (1572–1631) was born in England to a Catholic family. In 1601 he secretly married Anne Moore and, for this act, was sent to prison by her father. In 1615 he was ordained a priest and obtained a doctorate of divinity. In 1621 he was made Dean of St Paul's in London. He is recognized for the depth and beauty of his sermons as well as for his poetry.

*Rita Dove* served as Poet Laureate of the US from 1993–1995, the youngest poet and the first African-American to hold this honor. She won the Pulitzer Prize in 1987 for her verse cycle *Thomas and Beulah* and was elected a Chancellor of the Academy of American Poets in 2006. She is Commonwealth Professor of English at the University of Virginia.

*John L. Drost* has taught mathematics for over twenty years at Marshall University in Huntington, West Virginia, but he has been interested in prime numbers for much longer. His wife, Linda Hamilton, also teaches mathematics. He states that their married life is a good deal less "structured" than that of his poem's characters.

*Gerald Early* is Merle Kling Professor of Modern Letters at Washington University in St. Louis. He is an essayist, a critic of American culture, and author of six books, including *The Culture of Bruising,* which

won the 1994 National Book Critics Circle Award. His poetry collection is entitled *How the War in the Streets Is Won* (1995).

Enheduanna (c. 2285–2250 BC), high priestess in the city of Ur and daughter of King Sargon, was involved in mathematical practices and the composition of poems as part of her priestly duties. Her temple hymns are the earliest literary composition signed by their author. Her position of power as both en-priestess and royal princess may have enabled her to disregard convention and identify herself as author of her verses.

Cathryn Essinger is author of two poetry books: *A Desk in the Elephant House* (1998) and *My Dog Does Not Read Plato* (2004). She teaches writing at Edison Community College, Ohio. Her twin brother is a mathematician, which occasions conversations about the nature of language—literary, mathematical, and otherwise.

Jet Foncannon, formerly Jet Wimp, is Professor Emeritus of Mathematics at Drexel University. He has published in several areas of Applied Mathematics. He is also a poet and coeditor, with Ernest Robson, of the volume *Against Infinity: An Anthology of Contemporary Mathematical Poetry* (1979).

Robert Frost (1874–1963) is one of the most celebrated American poets. Many of his poems became beloved classics. His work was awarded the Pulitzer Prize four times during his lifetime. From 1958–1959 he served as Poet Laureate of the US. He lived most of his life in New England, where he farmed and taught poetry.

James Galvin is the author of six books of poems, a novel, and a work of nonfiction. Among his awards are a Guggenheim, an Ingram Merrill, and National Endowment of the Arts Fellowships. He works as a rancher in Wyoming part of each year and is on the permanent faculty of the University of Iowa's Writer Workshop for the rest of the year.

Martin Gardner is a popular recreational mathematician and science writer with a broad range of interests. Former author of the *Mathematical Games* column in *Scientific American*, he has published over 100 books, charming readers with his insights and humor. Gardner lives in Oklahoma, where occasional international conferences, known as "Gatherings for Gardner," are held in his honor.

Sarah Glaz is Professor of Mathematics at the University of Connecticut and author of *Commutative Coherent Rings* (1989; reissued in 2006) and other books and articles in commutative algebra. She also

has a lifelong interest in poetry, has served on the editorial board of *Ibis Review*, a literary magazine, and has published several of her poems and translations in periodicals.

*Douglas Goetsch* is author of six collections of poems, the most recent, *Your Whole Life*, winner of the 2007 Slipstream Press Poetry Contest. For 21 years he taught in New York City public school system and is now exploring other fields. He is founding editor of Jane Street Press.

*Thorold Gosset* (1869–1962) was an English lawyer and amateur mathematician who discovered and classified all semiregular polytopes in dimension four and higher. His work went largely unnoticed until it was rediscovered by Elte in 1912 and by Coxeter in 1973. The latter recounted the story and gave both Gosset and Elte credit for their work.

*Jennifer Granville* is a British writer, playwright, script editor, producer, and film actress. She teaches at Leeds Metropolitan University and the Ohio University School of Film. The poem "Cat in a Box" was inspired by the BIRS Workshop on Mathematics and Creative Writing, April 2004, which she attended with her brother, who is a mathematician.

*Mick Greer* divides his time between translating and teaching English at Lisbon University, Portugal. His research field is James Joyce and theatre.

*Emily Grosholz* is author of four books of poetry, most recently *The Abacus of Years* (2002), and advisory editor for *The Hudson Review*. She is Professor of Philosophy at Penn State University and author of *Representation and Productive Ambiguity in Mathematics and the Sciences* (2007).

*JoAnne Growney* has taught mathematics at Bloomsburg University, where she began to write poetry—first about math and then about everything. In 2001 she edited *Numbers and Faces*, a math-poetry collection published by the Humanistic Mathematics Network. In 2006 Paper Kite Press published her collection, *My Dance is Mathematics*.

*Bob Grumman*, a poet and critic living in Florida, has always enjoyed mathematics. Hence, he took two years of calculus while attending junior college in his thirties, although an English major. He made his first two math poems in the seventies, then none until the eighties. Now they are his specialty.

*Haipeng Guo* holds a PhD in computer science from Kansas State University and is currently a professor at United International College in

China. His research areas include Artificial Intelligence and theoretical computer science. His poem, "When a P-Man Loves an NP-Women," won the Computational Complexity Poetry Contest Award.

*Miriam Hammel* (1923–2001) was born in Austria and immigrated to Palestine in 1938. She studied English literature at Cambridge University but had a special interest in mathematics, especially the study of thermodynamics. She has published a novel, *A Seaman's Wife*, and poetry collections: *A String of Pearls* and *Miriam's Voices*.

*Israel Har* is an Israeli poet born in Poland. His family came to Palestine in 1933 and eventually settled in Jerusalem. He published several volumes of poetry, and served as editor in chief of the magazine *The Knapsack Library*. In 2002, over fifty of his poems appeared in the volume *Garden of Ropes Amid the Fire* .

*Robert Hass* is the author of several books of poetry, including 2008 Pulitzer Prize winner *Time and Materials*. Poet Laureate of the US from 1995–1997, he teaches English at the University of California, Berkeley. While still in college, he was an interested but not especially diligent student in the Great Books classes, where the curriculum included classical works of mathematics.

H. L. *Hix* directs the creative writing MFA program at the University of Wyoming. He is author of ten books, including five volumes of poetry. His most recent volume, *Chromatic* (2006), was a finalist for the National Book Award. About mathematics he says: "I dropped calculus during my freshman year of engineering school."

*Tony Hoagland* won the 2005 Mark Twain Award from the Poetry Foundation for humor in American poetry. His books of poems include *Sweet Ruin* (1992), which won the Brittingham Prize in Poetry, *Donkey Gospel* (1998), which received the James Laughlin Award, and *What Narcissism Means to Me* (2003). He teaches at the University of Houston.

*Roald Hoffmann* received the Nobel Prize in Chemistry in 1981 and is Professor of Humane Letters and a John A. Newman Professor of Chemistry at Cornell University. He also writes poetry, nonfiction, and plays. He published four books of poetry, the latest *Soliton* (2002). His writings have been translated into several languages.

*Jonathan Holden* is a University Distinguished Professor of English at Kansas State University and the first Poet Laureate of the state of Kansas. *Knowing* is his eighteenth book of poetry. His work has won

honors and awards, including the Vassar Miller Prize for *The Sublime*. His memoir, *Mama's Boys: A Double Life*, is forthcoming from Lewis-Clark Press.

*Langston Hughes* (1902–1967) was an important writer and thinker of the Harlem Renaissance. He wrote poetry, novels, short stories, plays, essays, children's books, and an autobiography. His writings depict, through personal experience, the common experience of the American blacks of his period.

*James Alston Hunter* (1902–1986), of Argentinean nationality, spent his life in England and Canada. After a first career as a finance officer in the British Royal Navy, in 1952 he moved to Canada, where he became a writer of a syndicated newspaper column, *Fun With Figures*. He was a prolific constructor of entertaining mathematical puzzles.

*Solomon Ibn Gabirol* (1021–1058) was a Spanish Hebrew poet and Neo-Platonic philosopher. Orphaned at an early age, he became a protégé of several patrons who admired his work. His best known poem is *A Kingly Crown*, a work which draws on his talents as poet and philosopher.

*Michael Kandel* is an American translator, author of science fiction books, and editor at Harcourt and at the Modern Language Association. He holds a PhD in Slavic Languages from Indiana University. Kandel is best known for the quality of his translations from Polish into English, in particular his translation of the works of Stanislaw Lem.

*Lisa Katz* is an Israeli poet and translator born in New York. *Reconstruction*, a book of her poetry translated into Hebrew, is forthcoming in Israel. Her work appeared recently in the US in the anthology *Illness in the Academy* (2007). Her translation of Agi Mishol's poetry, *Look There*, was published in 2006. She teaches at Hebrew University.

X. J. *Kennedy* is author of the 2007 collections *In a Prominent Bar in Secaucus: New & Selected Poems*, *Peeping Tom's Cabin: Comic Verse*, and numerous other poetry volumes, children's books, and textbooks. Among his awards is a Guggenheim Fellowship. He lives in Massachusetts with Dorothy M. Kennedy, his wife and coauthor of several books. A Kennedy son is a professor of mathematics education.

*James Kirkup* is a British poet, novelist, dramatist and translator. Among other honors, in the mid-1990s he won the Japanese Festival Foundation Prize for *A Book of Tanka* and the Scott Moncrieff Prize

for literary translation. His complete works have been published by Salzburg University Press.

A. S. *Kline* is a British writer and translator. He studied mathematics at the University of Manchester and was Chief Information Officer for a UK Company before turning full-time to literary interests. He has translated from Latin, ancient Greek, classical Chinese, and European languages, and continues his interest in mathematics and the sciences.

*Chana Kronfeld* is Professor of Hebrew and Comparative Literature at the University of California, Berkeley, where her work focuses on literary history and translation studies. Her volume *On the Margins of Modernism* (1996) received the MLA Scaglione Prize. She cotranslated Yehuda Amichai's book *Open Closed Open*.

*Stanley Kunitz* (1905–2006) is one of the most distinguished and loved American poets. Among his many honors are a Pulitzer Prize, a National Medal of the Arts, two terms as Poet Laureate of the US, and a Robert Frost Medal. He taught for 22 years at Columbia University, dividing his time between New York City and Cape Cod, Massachusetts.

*Bob Kurosaka* retired from teaching college-level mathematics in 2001. He published two science fiction stories, "Those Who Can, Do," and "A Lot to Learn," both reprinted in several anthologies; and a few limericks appearing on a number of websites. His parents were Japanese, but Bob, born in Lake George, NY, characterizes himself as American as sushi pie.

*Jacqueline Lapidus* currently lives in Boston. Born in New York City, she lived for many years in France, where she was active in international feminist groups. Among her poetry publications are the volumes *Ready to Survive* (1975), *Starting Over* (1977), and *Ultimate Conspiracy* (1988).

*Philip Larkin* (1922–1985) was one of the finest English poets of the twentieth century. Following the publication of his collection of poems *The Whitsun Weddings*, he was awarded the Queen's Gold Medal for Poetry. He was educated at Oxford and worked most of his life as a librarian at the University of Hull.

*Tom Lehrer* is a singer, songwriter, and mathematician. His musical career flourished while he was a graduate student of mathematics at Harvard University. Since 1970 he has taught mathematics and musi-

cal theatre, with only occasional forays into writing. In 1972 he joined the faculty of University of California, Santa Cruz.

*Stanislaw Lem* (1921–2006) was a Polish science fiction writer whose works were translated into forty-one languages. Of half-Jewish origin, he survived the Nazi occupation of Poland in WWII by working as a car mechanic with false papers. After WWII, in spite of roadblocks set up by the communist regime, he managed to publish a substantial body of work.

*Gabriel Levin* is a poet and translator who lives in Jerusalem. His third collection of poetry, *The Maltese Dreambook*, is forthcoming from Anvil Press. His translations include Yehuda Halevi's poetry, *Poems from the Diwan* (2002), and, with Peter Cole and Yahiya Hijazi, Taha Muhammad Ali's poetry, *So What: New & Selected Poems* (2006).

*John Stephen Lew* (1934–2006) was an applied mathematician who spent most of his professional career at IBM. He had a serious interest in poetry. Nathaniel, his son, says: "Already a versifier in his youth, my father advanced his avocation while in the US Air Force by attempting to read all the poetry available in Albuquerque, NM."

*Bernard Lewis*, Cleveland E. Dodge Professor Emeritus of Near Eastern Studies at Princeton University, is a well-known scholar in the history of the Islam, especially the Ottoman Empire and the modern conflict in the Middle East. He is the author of over thirty books and numerous articles and a translator of several volumes of poetry.

*Israel Lewis* is a retired engineer who took up creative writing late in life. He says: "As my work tended to the analytical I was well versed in certain branches of applied mathematics.... When I began to write creatively it was in fiction... and in due course finding my true métier in poetry. The math and science were there and didn't go away."

*Henry Lok* (1553–1608) was the son of Anne Vaughan Lok, the first woman to publish a sonnet sequence—which she appended to her book of Calvin's sermons. Henry was part of the outer Elizabethan court circle, working as an intelligencer for the Queen. He published two volumes of sonnets, *Sundry Christian Passions* (200 sonnets) and *Ecclesiastes* (320 sonnets).

*Marjorie Maddox*, Professor of English at Lock Haven University, is intrigued by issues of time and space and the intersections of boundaries. She is author of short stories and essays; eight volumes of poetry,

the most recent *Weeknights At The Cathedral,* 2006; and editor of anthologies. Her work has won a number of awards and prizes.

*Graça Margarido* is a professional translator working in Lisbon, Portugal.

*Edwin Markham* (1852–1940) was an American poet whose poem "The Man with a Hoe" became the impulse behind the social movement in poetry. He was raised on a ranch in California and attended school sporadically in his youth. Opposing his family's wishes, he enrolled in college and upon graduation embarked on a career as a writer.

*Andrew Marvell* (1621–1678) was an English poet whose involvement in politics kept his poetry from becoming known and appreciated till the twentieth century. After serving as Milton's Latin secretary, he held office in Cromwell's government. His poetry was compiled and published posthumously by his nephew in *Miscellaneous Poems.*

*Kaz Maslanka* is an engineering consultant, artist, and blogger in California. Showing an alternative to the tradition of mathematics as the language of science, Maslanka views mathematics as a language of art and works to stretch the boundaries of mathematical poetry. To learn more, visit: www.kazmaslanka.com.

*Harry Mathews,* author of several novels and collections of poetry, lived for many years in Europe and now divides his time between Paris and Key West. Elected to the OULIPO in 1972, he began intermittently writing works involving mathematical structure. He is coeditor of *The Oulipo Compendium* (2005).

*James Clerk Maxwell* (1831–1879) was a Scottish mathematician and physicist famous for formulating the equations, known as *Maxwell Equations,* which unify the basic laws of electricity and magnetism. He is considered a forefather of modern physics. He was a great lover of poetry and published a number of poems himself.

*John Ward McClellan* (1908–1986) was an American lithographic artist whose work can be found in a number of major museums. He settled in Woodstock, NY, in late 1930s. His lifelong interest in recreational mathematics and wordplay prompted a lively exchange of letters with Martin Gardner. This interest is also reflected in his art.

*Peter Meinke* has published fifteen books of poetry, seven in the prestigious Pitt Poetry Series, the latest being *The Contracted World.* A crystallographer daughter and a son who is director of Medicinal Chemis-

try at Merck provoke and inspire his scientific and mathematical poems and stories. His most recent book, *Unheard Music*, is a collection of short fiction.

*Dunya Mikhail* was born in Iraq, where she completed a degree in English literature. She published four books of poetry in Arabic. Her first work to appear in English is *The War Works Hard*, translated by Elizabeth Winslow, which was picked by the New York Public Library as one of 2005 Books to Remember.

*Wilmer Mills* holds a BA in English literature and an MA in theology. He lives with his wife and children in Tennessee, where he works as a carpenter. He published his poetry in *Right as Rain* (1999), *Light for the Orphans* (2002; reprinted 2004), anthologies, and numerous magazines.

*Agi Mishol* is an Israeli poet born in Transylvania, Romania, to a Hungarian family. She is the author of a dozen poetry books and the winner of numerous awards, most recently the 2007 Dolitsky Prize. Her poetry has been translated into several languages, including English (*Look There*, 2006). Mishol, currently Poet-in-Residence at the Hebrew University, lives on a farm and lectures at Alma College.

*Wendy Mnookin* is a poet living in Newton, Massachusetts. She is author of several collections of poetry, most recently *What He Took*, recipient of the 2002 New England Poetry Prize, and *The Moon Makes Its Own Plea* (forthcoming in 2008), both published by BOA Editions. More about her work at www.wendymnookin.com.

*Kyoko Mori* is author of three novels: *Shizuko's Daughter*, *One Bird*, and *Stone Field, True Arrow*; two nonfiction books: *The Dream of Water* and *Polite Lies*; and a collection of poems: *Fallout*. She teaches creative writing at George Mason University. Her awards include Best Novel of the Year from the Wisconsin Council of Writers.

*Randall Munroe* is a Christopher Newport University graduate with a degree in physics and formerly worked on robots for NASA. He is currently an Internet artist and cartoonist, maintaining and showing his work at xkcd.com, one of the most popular webcomics on the Internet.

*Howard Nemerov* (1920–1991) was a US Air Force combat pilot during World War II and had a continuing interest in the stars and navigation. He served twice as US Poet Laureate. *The Collected Poems*

*of Howard Nemerov* (1977) won the National Book Award and the Pulitzer Prize. Novelist as well as poet, Nemerov was a Distinguished Professor of English at Washington University.

*Pablo Neruda* (1904–1973) is the pen name of the Chilean writer and political activist Ricardo Eliecer Neftali Reyes Basoalto. He received the Nobel Prize for Literature (1971) and is considered one of the greatest poets of the twentieth century. He held several diplomatic positions until a change of government resulted in years of exile in Argentina.

*Susanna Nied* is a former instructor of English and comparative literature at San Diego State University. Her awards for translations of Inger Christensen's poetry include a 1982 American-Scandinavian PEN Translation Prize for Poetry, for *alphabet*, and a 2007 Harold Morton Landon Translation Award, for *it*.

*Deanna Nikaido*, a native of Denver, graduated from Art Center College of Design with a degree in illustration, but her passion is for writing. She has been featured at various venues, including reading festivals and museum events. Her poetry volume *Vibrating with Silence* was published by The Writer's Lair Books, Inc. in 2007. She resides in Baltimore.

*Katharine O'Brien* (1922–1998) was a mathematician, musician, and poet. She received a PhD in mathematics from Brown University and taught mathematics at Smith College, College of New Rochelle, and Deering High School. She published poems in numerous magazines and in *Excavation and Other Verse* (1967).

*Ewald Osers* translated nearly 150 collections of poetry from German and Czech into English, and Seamus Heaney's poetry into Czech. Osers speaks of translation using a line from Alexander Pope: "I lisped in numbers and the numbers came." Of Cibulka's "Mathematics," he says that neither poet nor translator has any particular connection with mathematics.

*Bill Parry* (1934–2006) was Professor of Mathematics at the University of Warwick. He is known for his work in the area of ergodic theory. In 1984 he was elected Fellow of the Royal Mathematical Society. After his retirement he became increasingly interested in poetry, and several of his poems have been published.

*Margaret Sayers Peden* is a scholar and translator of Spanish works who has translated over thirty-five books by major Latin American writers such as Allende, Vargas Llosa, Neruda, and Paz. Her work has

received numerous honors and awards, including the Gregory Kolovacas Award from PEN. She is Professor Emerita at the University of Missouri.

*Raymond Queneau* (1903–1976) was a French writer and publisher. He is known for his poetry and for his role in OULIPO, Ouvrior de Littérature Potentielle, a group of writers and mathematicians interested in constrained and experimental writing. His novel *Zazie Dans Le Métro*, made into a film, became a classic of the absurdist school of cinema.

*Bin Ramke* is author of nine collections of poetry, most recently *Tendril* (2007). He edits *The Denver Quarterly* and teaches at the University of Denver and, on occasion, at the School of the Art Institute of Chicago. While still in high school he studied one summer with topologist R. L. Moore at the University of Texas.

*William J. Macquorn Rankine* (1820–1872) was a Scottish engineer who made contributions to the science of thermodynamics. He was the author of a number of popular engineering manuals. His interests were wide ranging and included botany, number theory, singing, playing and composing music, and the writing of humorous poetry.

*Len Roberts* (1947–2007) was the author of nine books of poetry, most recently *The Disappearing Trick* (2007). His work has won numerous awards, including a Guggenheim Fellowship, selection for the National Poetry Series for his volume *Black Wings*, inclusion in *The Best American Poetry*, and the Pushcart Prize.

*Pattiann Rogers* has published ten books of poetry, most recently *Generations* (2004). She is recipient of numerous honors and awards, including Guggenheim and Lannan Fellowships, and four Pushcart Prizes. She is not a mathematician but is fascinated by the subject and its ability to reveal the workings of the physical world.

*Becky Dennison Sakellariou*, born in New England, has spent most of her adult life in Greece. Her poems appeared in the US and Greece. The poem in this collection came out of an advanced math textbook that she edited for its English. She says: "I understood none of the content, but loved the terms so much that...I 'invented' other meanings for them and wrote a love poem!"

*Eryk Salvaggio* is a new media artist and writer. He is the author of *Birds in Mid-Flight*, a collection of poems written as Scout Thompson (2001). Math was the weak point of his GPA and SAT scores until he

discovered the joy of computer programming at the age of 26. He lives in New England.

*Carl Sandburg* (1878–1967) was an American writer and historian. He won the Pulitzer Prize twice, once for the biography *Abraham Lincoln: The War Years*, and again for his *Complete Poems*. The son of poor Swedish immigrants, he was encouraged to embark on a writing career by his college professor, Phillip Green Wright, who recognized his talent.

*Maureen Seaton*, poet and hypermedia artist, received the Audre Lorde Award for her poetry collection *Venus Examines Her Breast*. Other awards include Lambda and Pushcart Prizes. Her connection to numbers began when she realized that they mysteriously fill the spaces between words and that they are beautiful.

*Ed Seykota* is a commodity trader who holds a degree in electrical engineering and industrial management from MIT. He resides in Nevada. His views on trading, his ideas on the Bernoulli principle in physics, and the complete fractal poem "Borderline," an excerpt of which is reproduced in this volume, appear on his website, www.seykota.com.

*Shel Silverstein* (1930–1999) was a composer, artist, and author of numerous books of poetry and prose for young readers. His prose includes a number of modern classics, and his poetry includes the award-winning collection of children's poems *Where the Sidewalk Ends* (1974). His intelligent and quirky humor is beloved by children and adults alike.

*Åke W. Sjöberg* is Professor Emeritus at the University of Pennsylvania and founding director of the Sumerian Dictionary Project (1976–1996). A noted scholar of Sumerian culture, Sjöberg, in collaboration with E. Bergmann, reconstructed and translated the temple hymns of Enheduanna (c. 2285–2250 BC) from fragments of cuneiform inscribed clay tablets.

*David R. Slavitt* is on the faculty of Yale University and Bennington College. He is author of more than 70 works of fiction, poetry, drama, and translation, and recipient of numerous honors and awards. His most recent translation work is *Sonnets of Love and Death, by Jean de Sponde* (2001).

*Young Smith* has received fellowships from the National Endowment for the Arts and the Kentucky Arts Council. His poems appeared in *Poetry, Beloit Poetry Journal, The Iowa Review, Pleiades, Crazyhorse, The*

*Harvard Divinity Bulletin*, and other publications. He is Assistant Professor of English at Eastern Kentucky University.

*Elizabeth Anne Socolow* is a writer and teacher whose love of mathematics originated in high school where she learned about asymptotes and the Folium of Descartes. She is the winner of the Barnard Poetry Prize for *Laughing at Gravity: Conversations with Isaac Newton* (1988) and for *Between Silences and Praise* (2006).

*Frederick Soddy* (1877–1956) was a distinguished English chemist and Professor at Oxford University and was awarded the Nobel Prize in Chemistry in 1921. His interest was in physical chemistry and radioactive materials. He coined the name *isotope* and is best known for his work on isotopes.

*King Solomon* (c. 1000–928 BC), the greatest of all ancient Hebrew Kings, reigned from 970 to 928 BC. The Bible attributes to Solomon the authorship of three books: *The Book of Proverbs*, a collection of fables and aphorisms; *Ecclesiastes*, a book of philosophy and contemplation; and *The Song of Songs*, a prose poem of erotic love.

*Helen Spalding* (1920–1991) was a British poet and novelist, most likely of Scottish origins. She is the author of a poetry collection, *What Images Return* (1947) and a nonfiction book, jointly with mathematician Hyman Levy, *Literature for an Age of Science* (1952). Her latest known piece appeared in *The London Magazine*, 1961.

A. E. *Stallings* is an American poet residing in Greece. She published two collections of poetry, *Archaic Smile*, winner of the Richard Wilbur Award, and *Hapax*. *The Nature of Things*, her verse translation of the Roman philosopher Lucretius, is out from Penguin. "Sine qua Non" was written for her father, who taught statistics at Georgia State University.

C. K. *Stead* is one of New Zealand's foremost literary figures. In 1968 he took early retirement from his position as Professor of English at the University of Auckland to write full time. His recent poetry book is *The Black River* (2007). He is a Fellow of the Royal Society of Literature, and recipient of his country's highest award, the Order of New Zealand.

*Mark Strand* teaches English at Columbia University. He is the author of 12 books of poetry, including Pulitzer Prize winner *A Blizzard of One* and a number of volumes of prose, translations, and children's

literature. His many honors include a MacArthur Foundation Award. From 1990–1991, Strand served as Poet Laureate of the US.

*Stephanie Strickland* is author of *Zone: Zero*, her fifth collection of poetry (forthcoming in 2008). Her double book from Penguin, *V: WaveSon.nets/Losing L'una*, has a web component. She studied mathematics at Harvard, and her next manuscript, *Huracan's Harp*, has many math poems, as does *True North*. Her latest collaborative hypermedia work is *slippingglimpse*. Visit: stephaniestrickland.com.

*Michael Stueben* has been trying, for over a decade, to combine the Moore Method of teaching mathematics with high school instruction at Thomas Jefferson H.S. for Science and Technology in Alexandria, Virginia. He published *Twenty Years Before the Blackboard* (1998), a book providing a delightful view of a math teacher's experience.

*Carolyn L. Tipton* holds a PhD in comparative literature from the University of California, Berkeley, where she now teaches. Awards and honors for her work include a National Endowment for the Humanities Fellowship. Her book *To Painting: Poems by Rafael Alberti* won the American Literary Translators Association National Translation Award.

*John Updike* is an internationally acclaimed American writer, author of 22 books and numerous essays, short stories and poems. His many awards include two Pulitzer Prizes. He was born in Pennsylvania and educated at Harvard and Oxford. His father was a high school mathematics teacher. He is father to four children, grandfather to seven.

*Amy Uyematsu* is a sansei (third generation Japanese-American) poet, anthology editor, and high school mathematics teacher from Los Angeles. She has published three collections of poetry, most recently *Stone Bow Prayer* (2005). Uyematsu is a recipient of the 1992 Nicholas Roerich Poetry Prize.

*Alissa Valles* studied at the school of Slavonic & East European Studies in London and works as a writer, editor, and translator from Russian, Polish, and Dutch. Her poems have appeared in *The Antioch Review*, *TriQuarterly*, *Ploughshares*, *Verse*, *Poetry*, and elsewhere. Her poetry collection *Orphan Fire* is forthcoming from Four Way Books.

*Ramón López Velarde* (1888–1921), a Mexican poet, studied law and worked as a judge, journalist, and civil servant. During his short life he published two volumes of poetry. His third volume, *El son del Corazon*, was published posthumously in 1932. His poetry did not gain recogni-

tion till after his death, when he became celebrated as Mexico's national poet.

*John Vieira*, poet and visual artist, has published ten chapbooks, the most recent of which is *Long Shot: An Autobiography of a Distant Friend* (2005). His poems have appeared in periodicals in the US and Canada. Numbers and mathematical symbols—in their brevity of expression and even in their beauty of form—have always stimulated him.

*Andrei Voznesensky* is a Russian poet. He suffered persecution during the communist regime, but with the changes in Russian politics his unique poetic voice brought him national and international recognition. In 1978 he was awarded the USSR State Prize for poetry. He is a honorary member the American Academy of Arts and Letters.

*Brenda Walker* has had a long and varied career divided between the arts and education. Her work has received a number of awards, including a 2002 Romanian Writers' Union Grand Translation Prize for her translation, with Stelian Apostolescu, of the poetry of Lucian Blaga. They have also cotranslated the poetry of George Bacovia (2006).

*Helen M. Walker* (1891–1983) had a distinguished career as statistician and educator. She obtained a doctorate from Columbia University, where she later became a professor. She held the positions of President of the American Statistical Association (1944) and President of the American Educational Research Association (1949–1950).

*William Wall* is an Irish writer and translator. He received a Patrick Kavanagh Award for Poetry for *Mathematics and Other Poems* (1997) and several other honors and awards for his fiction. He has published two volumes of poetry, four novels, and a collection of short stories. His work has been translated into several languages.

*Ronald Wallace* is Professor of English at the University of Wisconsin, where he also serves as codirector of the Creative Writing Program and editor of University of Wisconsin Press Poetry Series. He published 12 volumes of poetry, short stories, and criticism. He is recipient of numerous honors, including a Lifetime Achievement Award.

*Ellen Wehle* owes her love of numbers to her father, an engineer. She received an MFA in poetry from Florida International University in 1996 and has taught literature and creative writings at various colleges. For three years she was poetry editor at Agni. Her poems have been published in Canada, Australia, Europe and the US. She lives in Salem, Massachusetts.

*Joyce S. Weiner* holds an MBA and a PhD in romance languages from the University of Pennsylvania. She taught French for ten years at the University of Pennsylvania and Drexel University, during which time she did numerous translations. She is currently a clinical publications manager in the pharmaceutical industry.

*John Wheelwright* (1897–1940) was a poet, rebel Boston Brahmin, heretical Christian, and political figure in the American Marxist movement. His poetry is as experimental and eccentric as his short life. During his lifetime he published one collection of poems. Other poems were published posthumously in *Collected Poems* (1972).

*Williams Carlos Williams* (1883–1963) was educated in Switzerland, France, and the US. His distinguished literary career includes publications of short stories, plays, novels, translations, an autobiography and numerous volumes of poetry. In 1952 he was elected US Poet Laureate. He worked as a pediatrician for over 40 years.

*Elizabeth Winslow* is an American writer and translator whose work appeared in numerous periodicals. Her translation of Dunya Mikhail's poetry from Arabic, *The War Works Hard* (2005), won the PEN prize for translation in 2004 and was short-listed for the Griffin Poetry Prize in 2006.

*Clarence R. Wylie, Jr.* (1911–1995) obtained a PhD in mathematics from Cornell University. In his long academic career he held faculty positions at Ohio State University, the University of Utah, and Furman University. He is author of several bestselling mathematics textbooks and a book of poems, *Strange Havoc* (1956).

*Yang Mu* is the pen name of Ching-Hsien Wang, Professor Emeritus of Comparative Literature at the University of Washington, Seattle. He is a highly respected essayist, poet, and literary critic. His poetry, translated into English by Michelle Yeh and Lawrence R. Smith, appeared in *No Trace of the Gardener: Poems of Yang Mu* (1998).

*Michelle Yeh* holds a PhD in comparative literature from the University of Southern California and is currently Professor of East Asian Languages and Cultures at the University of California, Davis. Among her numerous studies and translations are *An Anthology of Modern Chinese Poetry* and *No Trace of the Gardener: Poems of Yang Mu.*

# *About the Mathematicians Appearing in the Poems*

We extend grateful acknowledgment to The Mac-Tutor History of Mathematics Archives, University of St. Andrews, Scotland, published by John J. O'Connor and Edmund F. Robertson (http://www-history.mcs.st-and.ac.uk/~history), which served as source for much of the historical information in this section.

*James W. Alexander* (1888–1971) was an American mathematician who was elected to the American Academy of Science for important contributions to the field of topology. He was educated in Princeton, Paris, and Bologna and worked most of his life at the Institute for Advanced Study, Princeton. (178)

*Archimedes* (287–212 BC) of Syracuse was the greatest mathematician of the ancient world. His calculations of volumes and areas anticipated modern calculus, and his work on enumeration anticipated modern combinatorics. His contribution to physics includes the famous theorem (*Eureka!*) that gives the weight of a body immersed in liquid. (17, 75)

*Alain Aspect* is a French physicist who made important contributions to quantum mechanics theory. He is Professor at École Polytechnique and a Senior Researcher at CNRS (Centre National de la Recherche Scientifique). Aspect was elected member of the French Academy of Science and in 2005 was awarded the CNRS Gold Medal. (23)

*Stefan Banach* (1892–1945) was a Polish mathematician, considered the founder of functional analysis. A professor at Lvov Technical University, Banach's interests extended to include mathematics education, editorship of journals and monograph series, and presidency of the Polish Mathematical Society. (38)

*Jakob Bernoulli* (1654–1705) See Contributors' Notes. (38, 130)

*George Boole* (1815–1864) was a British mathematician known for establishing a connection between logic and algebra. Forced from an early age to support his family, he ran his own school while teaching himself mathematics. His research brought him recognition and eventual appointment as a professor at Queen's College, Cork. (38)

*Georg Cantor* (1845–1918) was a Russian mathematician who lived most of his life in Germany. He made major contributions to set theory and to the study of enumeration by introducing cardinal numbers and exploring the various types of infinity. (138, 141)

*Augustin-Louis Cauchy* (1789–1857) was one of the nintheenth-century French mathematicians who founded the mathematical study of analysis. His contributions to this area had applications to physics. The turbulent political climate of his day contributed to his frequent changes of teaching positions. He taught in France, Switzerland, Italy, and Czechoslovakia. (38)

*Elwin Christoffel* (1829–1900) was a German mathematician known for his work in mathematical analysis, in which he was a follower of Dirichlet and Riemann. (38)

*Nicolaus Copernicus* (1473–1543) was a Polish astronomer and mathematician, physician, and translator of Greek poetry into Latin. He is renowned as a proponent of the views that the earth spins daily about its axis and revolves yearly around the sun, views that were in opposition to the official church vision of an earth-centered universe. (151)

*René Descartes* (1596–1650) was a French philosopher and mathematician. His work, *La Gèomèrie,* introduced Cartesian geometry and emphasized the role of algebra in formulating and solving geometric problems. His work had a major impact on both philosophy and mathematics. (24, 188)

*Albert Einstein* (1879–1955) was a celebrated physicist whose theories shaped the modern vision of physical reality. Among his numerous honors are the Nobel Prize (1921) and the Copley Medal (1925). Born in Germany, he fled the Nazi regime in 1932 and settled for the rest of his life at the Institute for Advanced Study in Princeton. (23)

*Euclid* (325–265 BC) was a Greek Mathematician who lived in Alexandria, a Greek colony situated in what is today Egypt. Euclid is the author of *The Elements,* a mathematical treatise which sets an axiomatic foundation for geometry. (27, 65, 87, 90, 92, 136, 151, 176, 188)

*Leonhard Euler* (1707–1783), a Swiss national who lived most of his life in St. Petersburg, Russia, is one of the greatest and most diverse mathematicians of the eighteenth century. His late marriage produced thirteen children, supporting his claim that he made some of his greatest mathematical discoveries while holding a baby in his arms. (38)

*Pierre de Fermat* (1601–1665), a French lawyer and mathematician, is best known for his work in number theory, in particular *Fermat's Last Theorem*, a conjecture that baffled mathematicians for over 300 years. *Fermat's Last Theorem* was proved in 1994 by Andrew Wiles. Fermat also made important contributions to the foundations of calculus. (97, 138, 146)

*Leonardo Fibonacci* (1170–1250) traveled extensively in his youth to Egypt, Syria, and Greece, where he was taught "Arabic" mathematics. On coming home to Italy, he wrote *Liber Abaci*, the book that introduced Europe to the Hindu-Arabic numeral system. The famous series of numbers known as Fibonacci Numbers first appeared in *Liber Abaci*. (7, 14, 75, 85)

*Joseph Fourier* (1768–1830) was a French mathematical physicist, who studied the theory of heat conduction, working with differential equations and infinite series. He was a professor at École Polytechniques. After Napoleon's defeat he was appointed secretary of the Academy of Sciences. (38)

*Carl Friedrich Gauss* (1777–1855) was a German mathematician who worked in a wide range of fields of mathematics and physics, including differential geometry and astronomy. His work had a substantial impact on each of these areas. He is considered one of the most influential nineteenth-century mathematicians. (27, 75, 155)

*J. Willard Gibbs* (1839–1903) was a prominent American physicist, chemist, and mathematician. He is well known for his contributions to chemical thermodynamics and physical chemistry and as the inventor of vector analysis. A Professor at Yale University, in 1901 he was awarded the Copley Medal. (191)

*Ronald Graham* is an American mathematician who has done important work in a number of areas, including computational geometry. From 1993–1994 he served as president of The American Mathematical Society and is currently an Irwin and Joan Jacobs Professor at the University of California, San Diego. (150)

*Stephen Hawking* is a British theoretical physicist who holds the post of Lucasian Professor of Mathematics at the University of Cambridge. He is renowned for his work in the fields of cosmology and quantum gravity and for his popular science bestsellers, in particular *A Brief History of Time*. (31)

*Werner Heisenberg* (1901–1976) was a German mathematician, best known for the *Uncertainty Principle*, discovered in 1927, which states that

the values of the position and momentum of a particle, when measured simultaneously, necessarily contain errors of certain magnitude. These errors, generally negligible, become critical when studying atom-size entities. (23)

*David Hilbert* (1862–1943), a German national, was one of the most influential mathematicians of the twentieth century. His work in geometry had the greatest impact on that subject since Euclid. At the 2nd International Congress of Mathematics, he proposed 23 problems, still considered to be some of the most significant questions in mathematics. (38, 176)

*Carl Jacobi* (1804–1951) was a Prussian mathematician who lived and worked in what is present-day Germany. He made serious contributions to the areas of elliptic functions, partial differential equations, and dynamics. (51)

*Johannes Kepler* (1804–1951) was a German mathematician and astronomer who is best known for discovering the three laws of planetary motion and advancing the cause of the heliocentric astronomy through exact calculations of astronomical tables. He also made important contributions to the theory of optics and to the area of solid geometry (90)

*Helge von Koch* (1870–1924) was a Swedish mathematician best known for a fractal curve resembling a snowflake and bearing his name. *Koch's curve* first appeared in a 1906 article; it is an example of a continuous function that is nowhere differentiable. (141)

*Gottfried Leibniz* (1646–1716) was a German mathematician who proved the *Fundamental Theorem of Calculus* at the same time as, and independently of, Isaac Newton. He is considered one of the forefathers of calculus and, therefore, of modern mathematics. He is also well known as a philosopher. (153)

*Thomas Malthus* (1766–1834) was a British economist best known for his work on population growth. He started his career as a country parson with an interest in mathematics and social philosophy. Publication of his essays on population growth led to his appointment as Britain's first professor in political economy. (66)

*Benoit Mandelbrot*, born in Poland, was educated and has lived in France and the US. He has worked at IBM and is currently Sterling Professor at Yale University. His work in fractal geometry is largely responsible for the recognition of the role fractals play in chaotic

systems, as well as for the computer-driven explorations of fractals' artistic beauty. (141, 186)

*Andrei Markov* (1856–1922) was a Russian mathematician known for his work in probability. From humble origins, he eventually gained fame as a mathematician and a position as a professor at St. Petersburg University. Markov had an interest in poetry and applied his mathematics to study chains of vowels in Aleksandr Pushkin's poetry. (38)

*James Clerk Maxwell* (1831–1879) See Contributors' Notes. (169, 191)

*Gregor Mendel* (1822–1884) was an Austrian Augustinian monk and plant biologist whose work on heredity, based on his experiments with pea plants, became the basis of the modern theory of genetics. (79)

*August Möbius* (1790–1868) was a German mathematician and physicist with interests in topology, astronomy, and mechanics. He is best known for the surface that bears his name, the *Möbius strip*, a two-dimensional surface with only one side. (166)

*John Napier* (1550–1617) was a Scottish scholar with a wide range of interests, primarily in agriculture and theology. As a mathematician he is best known for the invention of logarithms. His contributions to mathematics include a number of other calculating devices and a few formulas in geometry and trigonometry. (73)

*Isaac Newton* (1643–1727) was a celebrated British mathematician and one of the greatest scientists the world has ever known. He laid foundations for calculus and, with them, a foundation for modern mathematics. He also made fundamental contributions to physics, in particular to optics and celestial mechanics. (56, 75, 153)

*Emmy Noether* (1882–1935) was a German mathematician whose major contribution lies in abstract algebra, particularly in the structure of rings. Of Jewish origin, she fled Nazi Germany during WWII and immigrated to the United States, where she taught on a visiting professor position at Bryn Mawr College and lectured at the Institute for Advanced Study, Princeton. (158)

*Max Noether* (1844–1921) was a German mathematician who made contributions to algebraic geometry. He was the father of Emmy Noether, an algebraist with similar interests whose results generalized her father's. (158)

*Giuseppe Peano* (1858–1932) was an Italian mathematician considered one of the founders of mathematical logic. His work centered on the foundations of mathematics and the development of a formal symbolic language. (141)

*Pythagoras* (569–475 BC) was born in Samos, Greece, but settled in Crotona, Italy, where he founded a philosophical society whose aim was the study of mathematics. He made major contributions to the development of mathematics, astronomy, and the theory of music. The Pythagoreans were the first to solidify the notion of rigorous proof. (108)

G. F. *Bernhard Riemann* (1826–1866) was a German mathematician whose ideas concerning geometry of space had a major impact on the development of modern theoretical physics. His work on prime numbers gave rise to *The Riemann Hypothesis*, which is one of the important unsolved problems of mathematics. (38)

*Jacques Roubaud* is a French mathematician, poet, novelist and translator. An Emeritus Professor of Mathematics at the University of Paris X and Professor of Poetics at the École des Hautes Étude en Sciences Sociales, he has been a member of OULIPO since 1966. Roubaud describes himself as "a manufacturer of mathematics and poetry." (44)

*Bertrand Russell* (1872–1970) was a Welsh mathematician and philosopher whose best known work, *Principia Mathematica*, investigates the foundations of mathematics. Before their resolution, his paradoxes had shaken his contemporaries' trust in the underlying principles of mathematical knowledge. (65)

*Erwin Schrödinger* (1887–1961), an Austrian physicist, won the Nobel Prize in 1933. *Schrödinger's cat* is a paradox inherent in a thought experiment: a cat enclosed in a box lives or dies according to whether a quantum event occurs. Until the box is opened, the universe with the dead cat in it and the universe with the live cat have parallel existence. (23)

*Waclaw Sierpinski* (1882–1969) was a Polish mathematician who made contributions to several areas of mathematics, including set theory, number theory, and topology. His work in topology led him to the discovery of the *Sierpinski curve*, an infinite curve that encloses a finite area. (141)

*John Strutt, Lord Rayleigh* (1842–1919) was a British scientist who worked on the theory of waves. Professor of Physics at Cambridge

University, he was recipient of many honors and awards, including the Nobel Prize for the discovery of the gas *argon*. (191)

*Yutaka Taniyama* (1927–1958) was a Japanese mathematician who worked in algebraic number theory. He is famous for a conjecture he made in 1955 named, in his honor, *The Taniyama Conjecture*, which became a major factor in Andrew Wiles' proof of *Fermat's Last Theorem*. (51)

*Thales* (624–547 BC), the earliest known Greek mathematician, lived in Miletus in Asia Minor and was by profession an engineer. None of his writings survived, but legends credit him with five theorems of elementary geometry and report that he accurately predicted the eclipse of the sun in 585 BC. (27)

*William Thomson, Lord Kelvin* (1824–1907) was a Scottish physicist, professor at Glasgow University, creator of the first physics laboratory, and inventor of the mirror galvanometer, a device that can read weak electrical signals with the aid of a mirror. (169)

*John Venn* (1834–1923) was a British mathematician who worked in the area of mathematical logic and wrote a small number of highly original and influential articles. He is best known for the diagrammatic way of representing sets and their unions and intersections as overlapping disks. (38, 40, 58)

*Ludwig Wittgenstein* (1889–1951) was a nineteenth-century Austrian philosopher. His first work, *The Tractus*, inspired by Bertrand Russell, contributed to the foundations of mathematics. His second work, *Philosophical Investigations*, inspired by the mathematician Frank Ramsey, contributed to the philosophy of language. (3, 32)

*Zeno* (490–425 BC) of Elea was a Greek philosopher known for posing paradoxes that challenged the logic of mathematical thought for many centuries. His most famous is the *Achilles and the Tortoise Paradox*. (10, 48, 137)

# Acknowledgments

We extend grateful acknowledgment to the authors and their families, translators, editors, literary estate agents, universities, and publishers who have given us permission to reprint these selections, and to all the sources that made the poems in this volume available to us.

*Diane Ackerman:* "Kismet III" from *Jaguar of Sweet Laughter* by Diane Ackerman, copyright © 1991 by Diane Ackerman. Used by permission of Random House, Inc.

*Rafael Alberti:* "Harold Lloyd, Student" from *Looking for Poetry, Poems by Carlos Drummond and Rafael Alberti and Songs from the Quechua,* translated by Mark Strand, Alfred A. Knopf, 2002, copyright © 2002 by Mark Strand. Used by permission of the translator. "To Divine Proportion" from *To Painting: Poems by Rafael Alberti,* translated by Carolyn L. Tipton, Evanston: Northwestern University Press, English translation copyright © 1997 by Carolyn L.Tipton. Used by permission of Northwestern University Press and the translator.

*Sandra Alcosser:* "My Number," copyright 1998 by Sandra Alcosser. Reprinted from *Except by Nature* with the permission of Graywolf Press, Saint Paul, Minnesota, and the author.

*Manuel Alegre:* "Polar Heart" (excerpt) from Instituto Camoes: www.instituto-camoes.pt/CVC/oceanoculturas/eng/15.html, translated by Mick Greer and Graça Margarido. Used by permission of the translators.

*Karren LaLonde Alenier:* "Dialectic of the Census Takers" first appeared in *Negative Capability,* Vol. IX, # 2, 1989. Used by permission of the author.

*J. W. Alexander:* The image of Alexander's Horned Sphere is from "An Example of a Simply Connected Surface Bounding a Region Which is Not Simply Connected," *Proceedings of the National Academy of Sciences* 10 (1924), pp 8–10, copyright © J. W. Alexander.

*Dante Alighieri:* "The Paradiso: Canto XXXIII" (lines 124–146) from *The Divine Comedy,* translated by John Ciardi. Copyright 1954, 1957, 1959, 1960, 1961, 1965, 1967, 1970 by the Ciardi Family Publishing Trust. Used by permission of W.W. Norton & Company, Inc.

*Yehuda Amichai:* Excerpt from "Israeli Travel: Otherness is All, Otherness is Love" in *Open Closed Open,* copyright © 2000 by Yehuda Amichai, English translation copyright © 2000 by Chana Bloch and Chana Kronfeld. Reprinted by permission of Harcourt, Inc., and Georges Borchardt, Inc., on behalf of Chana Bloch and Chana Kronfeld.

*Carl Andre:* "On the Sadness" has appeared in *Numerals 1924–1977,* by Rainer F. Crone in collaboration with the graduate seminar at Yale University, Yale University Art Gallery Catalog Exhibit, 1978. Used by permission of Melissa Kretschmer on behalf of the artist.

*Amy Quan Barry:* "If $\frac{dy}{dx} = \frac{4x^3 + x^2 - 12}{\sqrt{2x^2 - 9}}$, then" from *Asylum* by Amy Quan Barry © 2001. Reprinted by permission of the University of Pittsburgh Press and the author.

*Judith Baumel:* "Fibonacci" from *The Weight of Numbers,* Wesleyan University Press, copyright © 1988 by Judith Baumel. Used by permission of the author.

*Jakob Bernoulli:* "Treatise on Infinite Series" (excerpt), translated by Helen M. Walker, from *A Source Book in Mathematics,* by David Eugene Smith, Dover Publications, Inc., 1959.

*Bhaskaracharya:* Poem from "Lilavati," by Bhaskaracharya (1150), translated in *The Universal History of Numbers* by George Ifrah (translators: David Bellos, E. F. Harding, Sophie Wood and Ian Monk), published by Harvill (translation copyright © 1998 by The Harvill Press Ltd). Reprinted by permission of The Random House Group, Ltd.

*Linda Bierds:* "Counting: Gregor Mendel in the Prelacy" from *First Hand* by Linda Bierds, copyright © 2004 by Linda Bierds. Used by permission of Marion Wood Books, an imprint of G. P. Putnam, a division of Penguin Group (USA) Inc., and the author.

*Lucian Blaga:* "Tally" from *Complete Poetical Works of Lucian Blaga, 1895–1961,* translated by Stelian Apostolescu and Brenda Walker, copyright © 2001 by UNESCO/Center For Romanian Studies for the English translation. Used by permission of the translators.

*Ray Bobo:* "Give Me an Epsilon and I Will Treat It Well" from *The American Mathematical Monthly* 87, 1980. Used by permission of the author.

*David Brooks:* "Yes" will appear in *The Balcony,* by David Brooks, University of Queensland Press, 2008. Used by permission of the author.

*Elizabeth Barrett Browning:* "Sonnet XLIII: How do I love thee? Let me count the ways" from *Sonnets from the Portuguese* (1850), also appears in *Poet's Corner,* www.theotherpages.org/poems.

*Ann Calandro:* "Where the Kissing Never Stops" first appeared in *Against Infinity: An Anthology of Contemporary Mathematical Poetry,* eds. Ernest Robson and Jet Wimp, Primary Press, 1979. Used by permission of the author.

*Lewis Carroll:* "The Hunting of the Snark: The Beaver's Lesson" (excerpt) from *The Hunting of the Snark* (1876), also appears in *Poet's Corner,* www.theotherpages.org/poems.

*Catullus:* "Let's Live and Love: To Lesbia," translated by A .S. Kline, from www.tonykline.co.uk, copyright © by A. S. Kline. Used by permission of the translator.

*Siv Cedering:* "The Acrobat" from *Letters From a Floating World,* by Siv Cedering, University of Pittsburgh Press, copyright © 1984 by Siv Cedering. Used by permission of the author.

*Robin Chapman:* "Nonlinear Function" from Chapman, Robin, & Sprott, J. C., *Images of a Complex World: The Art and Poetry of Chaos,* Singapore: World Scientific Publishing, copyright © 2005 by Robin Chapman and Julien Clinton Sprott. Used by permission of the author.

*Edward E. Chipman:* "God Loves a Curve" first appeared in *An Evening of Mathematical Poetry: January 10, 1992, National Joint Mathematical Meetings,* Baltimore, MD, also appeared in *The Humanistic Mathematical Network Journal* # 7, 1992. Used by permission of Susan Kline.

*Inger Christensen:* "Love" (excerpt) by Inger Christensen, translated by Susanna Nied, from *alphabet,* copyright © 1981, 2000 by Inger Christensen, translation copyright © 2000 by Susanna Nied. Reprinted by permission of New Directions Publishing Corp.

*Hanns Cibulka:* "Mathematics" from *Not Marked on the Map: Selected Poems by Hanns Cibulka,* translated by Ewald Osers, Aquila, Isle of Skye, 1985. Used by permission of the translator.

*Dan Clark:* "Practical Applications" from *Chicken Soup for the Teen Soul II,* edited by Jack Canfield, Mark Victor Hansen, and Kimberly Kirberger. Copyright © 1998 by Dan Clark. Reprinted with the permission of Heath Communications, Inc., www.hcbooks.com.

*Marion Deutsche Cohen:* "What Drove Me Into Math" from *Crossing the Equal Sign,* Plain View Press, TX, copyright © 2007 by Marion D. Cohen,

and "Scared and the Intermediate Value Theorem." Used by permission of the author.

Jennifer Crow: "Mathematics" from *Strange Horizons*, www.strangehorizons.com, copyright © 2002 by Jennifer Crow. Used by permission of the author.

Mary Cornish: "Numbers" from *Red Studio* by Mary Cornish, Oberlin College Press, © 2007. Reprinted by permission of Oberlin College Press.

Jonathan Coulton: "Mandelbrot Set," copyright © by Jonathan Coulton, from www.jonathancoulton.com/lyrics /mandelbrot-set. Used by permission of the author.

J. V. Cunningham: "Statistics" from *The Poems of J. V. Cunningham*, edited with an Introduction and Commentary by Tim Steele. Copyright © 1997. Reprinted by permission of Swallow Press/Ohio University Press, Athens, Ohio (www.ohioswallow.com).

Keith Allen Daniels: "Mathrimony" from *I Think Therefore Iamb*, by Keith Allen Daniels, Anamnesis Press, copyright © 2000 by Keith Allen Daniels. Used by permission of Toni Daniels.

Jon Davis: "The Accounting" from *Ploughshares*, Winter 2001–2002. Used by permission of the author.

Jean de Sponde: "Sonnet of Love XIII" from *Sonnets of Love and Death*, translated by David R. Slavitt, Evanston: Northwestern University Press, 2001, English translation copyright © 2001 by David Slavitt. Used by permission of Northwestern University Press.

E. P. Dempster: "A Mathematical Madness" (excerpt) from *A Book of Nonsense Verse*, ed. Langford Reed, G. P. Putnam's Sons, The Knickerbocker Press, 1926.

Theodore Deppe: "The Proof" from *Cape Clear: New and Selected Poems*, by Theodore Deppe, Salmon Publishing, Ltd., Ireland, copyright © 2002 by Theodore Deppe. Used by permission of the author.

Kathryn DeZur: "Fibonacci Numbers" from *Literary Mama: A Journal for the Maternally Inclined*, October 2005. Used by permission of the author.

James Dickey: "Math" from *The Whole Motion: Collected Poems, 1945–1992* (Wesleyan University Press, 1992) © 1992 by James Dickey and reprinted by permission of Wesleyan University Press.

Emily Dickinson: J26 "It's all I have to bring today," and J1484 "We shall find the cube of the rainbow." Reprinted by permission of the publishers and the Trustees of Amherst College from *The Poems of Emily Dickinson*, Thomas

H. Johnson, ed., Cambridge, Mass.: The Belknap Press of Harvard University Press, copyright © 1951, 1955, 1979, 1983, by the President and Fellows of Harvard College.

John Donne: "Love's Growth" and "The Computation" from *John Donne: The Complete English Poems*, edited by A. J. Smith, Penguin Books, 1971.

Rita Dove: "Geometry" from *The Yellow House on the Corner*, Carnegie Mellon University Press, © 1980 by Rita Dove. Reprinted by permission of the author. "Flash Cards" from *Grace Notes* by Rita Dove, copyright © 1989 by Rita Dove. Used by permission of the author and W. W. Norton & Company, Inc.

John L. Drost: "I Even Know of a Mathematician" from *The American Mathematical Monthly* 12, 2005. Used by permission of the author.

Gerald Early: "With Linnet and Rosalind on the Evening of the Equinox (or an Exposition on the Big Bang Theory)" from *The American Poetry Review*, September–October, 1995. Used by permission of the author.

Enheduanna: TH No.42 "Temple Hymn No. 42" (excerpt), from "The Collection of the Sumerian Temple Hymns," by Åke W. Sjöberg and E. Bergmann S. J., in: *Texts from Cuneiform Sources, Vol. III*, ed. A. Leo Oppenheim, Miguel Civil, Erica Reiner, Edmond Sollberger, J. J. Augustin Publisher, Locust Valley, NY, 1969. Used by permission of J. J. Augustin, Inc.

Cathryn Essinger: "My Dog Practices Geometry" first appeared in *Poetry*, February 2002, reprinted in *My Dog Does Not Read Plato*, Main Street Rag Publishing, copyright © 2004 by Cathryn Essinger. Used by permission of the author. "You Are Right" first appeared in *Poetry*, August 1992, and "The Mathematician, Counting" first appeared in *Negative Capability*, both reprinted in *A Desk in the Elephant House*, copyright © 1998 by Texas Tech University Press. Used by permission of the author and Texas Tech University Press.

Robert Frost: "Moon Compasses" from *The Poetry of Robert Frost* edited by Edward Connery Lathem. Copyright 1969 by Henry Holt and Company. Copyright 1936 by Robert Frost, copyright 1964 by Lesley Frost Ballantine. Reprinted by permission of Henry Holt and Company, LLC.

James Galvin: "Geometry Is the Mind of God" from *Resurrection Update: Collected Poems 1975–1997*, copyright © 1997 by James Galvin. Reprinted with the permission of the author and Copper Canyon Press, www.coppercanyonpress.org.

Martin Gardner: The $\pi$ quatrain, appearing in the Introduction, is used by permission of the author.

*Sarah Glaz:* "Calculus" from *The Humanistic Mathematics Network Journal* 26, 2002, "Love Story," and "Late Afternoon at the Workshop on Commutative Rings," used by permission of the author.

*Douglas Goetsch:* "Counting" reprinted from *Nobody's Hell* © 1999 by Douglas Goetsch, by permission of Hanging Loose Press.

*Thorold Gosset:* "The Kiss Precise (Generalized)" reprinted by permission of Macmillan Publishers Ltd: *Nature*, January 9, 1937, copyright 1937 by Macmillan Publishers Ltd.

*Jennifer Granville:* "Cat in a Box" from *The American Mathematical Monthly*, November 2004. Used by permission of the author.

*Emily Grosholz:* "Proportions of the Heart" and "The Shape of Desire" first appeared in *New Virginia Review*, reprinted in *Eden*, John Hopkins University Press, copyright © 1992 by Emily Grosholz, and "The Ratio of Green" *from Shores and Headlands*, copyright © 1998 by Princeton University Press, copyright © 2008 by Emily Grosholz. Used by permission of the author.

*JoAnne Growney:* "My Dance is Mathematics" and "San Antonio, January, 1993" from *My Dance is Mathematics*, by JoAnne Growney, Paper Kite Press, copyright © 2006 by JoAnne Growney, and "Floating," from joannegrowney.com, used by permission of the author. Epigraph to "My Dance is Mathematics": excerpt from "Dirge without Music" by Edna St. Vincent Millay, copyright © 1928, 1955 by Edna St. Vincent Millay and Norma Millay Ellis. Used by permission of Elizabeth Barnett, Literary Executor, The Millay Society.

*Bob Grumman:* "Mathematiku No. 10" from www.geocities.com/Compreppoetica/. Used by permission of the author.

*Haipeng Guo:* "When a P-Man Loves an NP-Woman" first appeared, in a different version, in *Computational Complexity*, www.weblog.fortnow.com. Used by permission of the author.

*Miriam Hammel:* "Entropy" first appeared in *Voices Israel* 1996, reprinted in *Miriam's Voices*, 1.7.1923 – 6.8.2001, copyright © 2002 Ya'aqov J. Hammel & Sons. Used by permission of Ya'aqov Hammel.

*Israel Har:* "The Zero," translated by Gabriel Levin, from *International Poets' Festival, Jerusalem 1995*, ed. Helit Yeshurun, Keterpress Enterprises. Used by permission of the translator.

*Robert Hass:* "White of Forgetfulness, White of Safety" from *Time and Materials: Poems 1997–2005*, by Robert Hass, copyright © 2007 by Robert Hass. Reprinted by permission of HarperCollins Publishers and the author.

*H. L. Hix:* "Orders of Magnitude" (decimals 39, 40, 41, 50 and 64) from *Rational Numbers*, Truman State University Press, copyright © 2000 by Truman State University Press. Used by permission of Truman State University Press and the author.

*Tony Hoagland:* "Self-Improvement" from *Donkey Gospel*, Graywolf Press, copyright © 1998 by Tony Hoagland. Used by permission of the author.

*Roald Hoffman:* "Why Does Disorder Increase in the Same Direction of Time as That in Which the Universe Expands?" first appeared in The *Madison Review*, 1994, reprinted in *Imaginary Numbers*, ed. William Frucht, John Wiley & Sons, 1999. Used by permission of the author.

*Jonathan Holden:* "Sex and Mathematics" and "The Departure of an Alphabet" from *Knowing: New and Selected Poems* by William Holden, The University of Arkansas Press, copyright © 2000 by Jonathan Holden. Used by permission of the author.

*Langston Hughes:* "Addition [1]" from *The Collected Poems of Langston Hughes* by Langston Hughes, ed. by Arnold Rampersad with David Roessel, Associate Editor, copyright © 1994 by the Estate of Langston Hughes. Used by permission of Alfred A. Knopf, a division of Random House, Inc., and Harold Ober Associates.

*J. A. H. Hunter:* "A Lady and Her Lingerie" from *Fun with Figures*, by J. A. H. Hunter, Dover Publications, 1965.

*Solomon Ibn Gabirol:* "The Praises of God" (excerpt) from *The Kingly Crown*, by Solomon Ibn Gabirol, translated by Bernard Lewis, University of Notre Dame Press, copyright © 2003 by University of Notre Dame. Used by permission of the University of Notre Dame Press.

*X. J. Kennedy:* "Geometry," first published in *Poetry*, October–November, 2002, reprinted in *In a Prominent Bar in Secaucus: New & Selected Poems* by X. J. Kennedy, John Hopkins University Press, copyright © 2007 by X. J. Kennedy. Used by permission of the author.

*James Kirkup:* "How to Find the Longest Distance Between Two Points." Used by permission of the author.

*Stanley Kunitz:* "Geometry of Moods" © 1978 by Stanley Kunitz, from *The Poems of Stanley Kunitz 1928–1978*, An Atlantic Monthly Press Book, Brown and Co., 1979. Permission granted by Darhansoff, Verrill, Feldman, Literary Agents. All rights reserved.

*Bob Kurosaka:* "There Was a Young Maiden." Used by permission of the author.

*Jacqueline Lapidus*: "Several Hypotheses and a Proposition" first appeared in *Hanging Loose Magazine*, reprinted in *Starting Over*, by Jacqueline Lapidus, Out & Out Books, 1977, and in *Against Infinity: An Anthology of Contemporary Mathematical Poetry*, ed. Ernest Robson and Jet Wimp, Primary Press, 1979. Used by permission of the author.

*Philip Larkin*: "Counting" from *Collected Poems* by Philip Larkin, copyright © 1988, 2003 by the Estate of Philip Larkin. Reprinted by permission of Farrar, Straus and Giroux, LLC., and Faber & Faber, Ltd.

*Tom Lehrer*: "There's a Delta for Every Epsilon" from *The American Mathematical Monthly* 81, 1974. Used by permission of the author.

*Stanislaw Lem*: Poem excerpt from *The Cyberiad: Fables for the Cybernetic Age*, by Stanislaw Lem, English translation by Michael Kandel, copyright © 1974 by Harcourt, Inc. Reprinted by permission of the publisher.

*John Stephen Lew*: "The Comet," from *The Humanistic Mathematics Network Journal*, May, 1996. Used by permission of Nathaniel G. Lew.

*Israel Lewis*: "I Find My Faith in the Flatness of Space" from mysite.verizon.net/scribblew/flatness.htm. Used by permission of the author.

*Henry Lok*: "Square Poem in Honor of Elizabeth I" from *Sundry Christian Passions*, London 1597. Image used by permission of Houghton Library, Harvard University.

*Marjorie Maddox*: "Venn Diagrams" appeared in *Oak Review* (2001), *Prairie Schooner* (1990), and in *Perpendicular As I*, by Marjorie Maddox, Sandstone, copyright © 1995 by Marjorie Maddox. Used by permission of the author.

*Edwin Markham*: "Outwitted," from *Epigrams* (1901), also appears in *Poet's Corner*, www. theotherpages.org/poems.

*Andrew Marvell*: "The Definition of Love" from *Miscellaneous Poems* (1681), also appears in *Poet's Corner*, www. theotherpages.org/ poems.

*Kaz Maslanka*: "Sacrifice and Bliss," used by permission of the author.

*Harry Mathews*: "Shakespeare on Love" from "Mathews' Algorithm" in *OULIPO: A Primer of Potential Literature*, ed. Warren F. Motte, Jr., University of Nebraska Press, 1986, and "Safety in Numbers." Used by permission of the author.

*James Clerk Maxwell*: "A Lecture on Thomson's Galvanometer" from *Nature*, May 16, 1872, Macmillan Publishers, Ltd.

*John Ward McClellan:* "A Lady of 80" from "Gauss's Congruence Theory Was Mod As Early As 1801" by Martin Gardner, *Scientific American*, February 1981. Used by permission of Suzan McClellan Whiting.

*Peter Meinke:* "Distances" from *The Night Train and the Golden Bird*, by Peter Meinke, © 1977, and "3.141592..." from *Scars*, by Peter Meinke, © 1996. Reprinted by permission of University of Pittsburgh Press and the author.

*Dunya Mikhail:* "The Theory of Absence," translated by Elizabeth Winslow, from *The War Works Hard*, copyright © 1993, 1997, 2000, 2005 by Dunya Mikhail. Reprinted by permission of New Directions Publishing Corp., Carcanet Press, Limited, and the author.

*Wilmer Mills:* "An Equation for My Children" from *Poetry*, June 2003. Used by permission of the author.

*Agi Mishol:* "Geese," translated by Lisa Katz, from *PBS Online News Hour: Poetry Series*: www.pbs. org/newshour/indepth_coverage/entertainment/poetry/profiles/poets_mishol.html. Used by permission of the translator and the author.

*Wendy Mnookin:* "Math" from *What He Took*, copyright © 2002 by Wendy Mnookin. Reprinted with the permission of the author and BOA Editions, Ltd., www.boaeditions.org.

*Kyoko Mori:* "Barbie Says Math Is Hard" from *Fallout*, Tia Chucha Press, copyright © 1994 by Kyoko Mori. Used by permission of the author.

*Randall Munroe:* "Useless" from www.xkcd.com. Used by permission of the artist.

*Howard Nemerov:* "Figures of Thought" from *The Collected Poems of Howard Nemerov*, University of Chicago Press, copyright © 1977 by Howard Nemerov. Used by permission of Margaret Nemerov.

*Pablo Neruda:* "Ode to Numbers" from *Selected Odes of Pablo Neruda*, by Pablo Neruda, Margaret Sayers Peden, translator, © 1999 Fundacion Pablo Neruda, published by the University of California Press (pages 95, 97, 99) copyright © 1990 by The Regents of the University of California. Used by permission of the University of California Press and the translator.

*Deanna Nikaido:* "July 18, 2005" from *Beltway Poetry Quarterly*, Spring 2007. Used by permission of the author.

*Katharine O'Brien:* "Valentine" and "Undefined Terms" from *Excavation and Other Verse*, by Katharine O'Brien, The Antheoen Press, 1967. Used with

permission from Bowdoin College, Brunswick, Maine; Brown University, Providence, Rhode Island; and The University of Maine System.

*Bill Parry:* "Argument" and "Alexander's Horned Sphere" from *Notices of the American Mathematical Society* 54, March 2007, copyright © by Bill Parry's Estate. All rights reserved. Used by permission of David Epstein and Benita Parry.

*Raymond Queneau:* "Swigns" translated by Jet Foncannon and Joyce S. Weiner, from *Against Infinity, an Anthology of Contemporary Mathematical Poetry*, ed. Ernest Robson and Jet Wimp, Primary Press, 1979. Used by permission of the translators.

*Bin Ramke:* "Gravity & Levity" (excerpt) from *Conjunctions: 35, American Poetry: State of the Art, Fall 2000*. Used by permission of the author.

*William J. Macquorn Rankine:* "The Mathematician in Love" from *Songs and Fables*, by William J. Macquorn Rankine, Glasgow, James Maclehose, 1874, also appeared in The University of Toronto Libraries, Representative Poetry Online (rpo.library. utoronto.ca/display/index.cfm).

*Len Roberts:* "We Sat So Patient" first appeared in *Boulevard*, reprinted in *The Best American Poetry 1992*, ed. Charles Simic, Series ed. David Lehman, Macmillan Publishing Company, 1992. Used by permission of Nancy Roberts.

*Pattiann Rogers:* "On Your Imminent Departure: Considering the Relative Importance of Various Motions" from *Song of the World Becoming: New and Collected Poems 1981–2001*, by Pattiann Rogers, Milkweed Editions, copyright © 2001 by Pattiann Rogers. Used by permission of the author.

*Becky Dennison Sakellariou:* "Math Is Beautiful and So Are You" from *Beloit Poetry Journal*, Fall 2002. Used by permission of the author.

*Eryk Salvaggio:* "Five Poems about Zero" (excerpt). Used by permission of the author.

*Carl Sandburg:* "Number Man" from *The Complete Poems of Carl Sandburg*, copyright © 1970, 1969 by Lilian Steichen Sandburg, Trustee, and "Child Margaret" from *Cornhuskers*, by Carl Sandburg, copyright 1918 by Holt, Rinehart and Winston, and renewed 1946 by Carl Sandburg. Reprinted by permission of Harcourt, Inc.

*Maureen Seaton:* "The Freezing Point of the Universe" from *Little Ice Age*, by Maureen Seaton, Invisible City Press, copyright © 2000 by Maureen Seaton. Used by permission of the author and Invisible City Press.

*Ed Seykota:* "Borderline—A Fractal Poem" (excerpt) from www.seykota.com, copyright © 1986 by Ed Seykota. Used by permission of the author.

*Shel Silverstein:* "Smart" from *Where the Sidewalk Ends*, by Shel Silverstein, HarperCollins Publishers, 1974. Copyright © 2004 by Evil Eye Music, Inc. Reprinted with permission from the Estate of Shel Silverstein and HarperCollins Children's Books.

*Young Smith:* "She Considers the Dimensions of Her Soul" from *Poetry*, July 2003. Used by permission of the author.

*Elizabeth Anne Socolow:* "Asymptotes" from *Isotope*, Fall-Winter 2006. Used by permission of the author.

*Frederick Soddy:* "The Kiss Precise" reprinted by permission of Macmillan Publishers Ltd: *Nature*, June 20, 1936, copyright 1936 by Macmillan Publishers Ltd.

*King Solomon:* "The Song of Songs" (excerpt), from *The Song of Songs: A New Translation, Introduction and Commentary*, translators Chana Bloch and Ariel Bloch, Random House, 1995; University of California Press, 1998; Modern Library Classic, 2006. Copyright © 1995 by Ariel Bloch and Chana Bloch. Used by permission of Chana Bloch on behalf of the translators.

*Helen Spalding:* "Let Us Now Praise Prime Numbers" from *An Anthology of Modern Verse 1940–1960*, ed. E. Jennings, Methuen, 1961. Every effort has been made to locate copyright holders.

*A. E. Stallings:* "Sine Qua Non" from *Hapax*, Evanston: Tri Quarterly Books/ Northwestern University Press, copyright © 2006 by A. E. Stallings. Used by permission of Northwestern University Press and the author.

*C. K. Stead:* "Walking Westward" (excerpt) from *Straw Into Gold: Poems New and Selected*, by C. K. Stead, Auckland University Press, copyright © 1999 by C. K. Stead. Used by permission of the author.

*Mark Strand:* "The New Poetry Handbook" from *Darker, Poems* by Mark Strand, Atheneum, copyright © 1970 by Mark Strand. Used by permission of the author.

*Stephanie Strickland:* "Who Counts, Counts," "True North 1," and "Striving All My Life" from *True North*, by Stephanie Strickland, University of Notre Dame Press, copyright © 1997 by Stephanie Strickland. Used by permission of the author and University of Notre Dame Press.

*Michael Stueben:* "Valentine" from www.pen.k12.va.us/Div/Winchester/jhhs/math/poetry/mathpoet.html. Used by permission of the author.

Unknown Author: "Puzzle" from *Puzzles from Other Worlds: Fantastical Brain Teasers from Isaac Asimov's Science Fiction Magazine*, by Martin Gardner, A Vintage Original, 1984.

John Updike: "Midpoint, IV: The Play of Memory" (excerpt), from *Collected Poems 1953–1993*, by John Updike, copyright © 1993 by John Updike. Used by permission of Alfred A. Knopf, a division of Random House, Inc. and the author.

Amy Uyematsu: "The Meaning of Zero: a Love Poem" and "Unfinished Calculations" from *Stone Bow Prayer*, Copper Canyon Press, copyright © 2005 by Amy Uyematsu. Used by permission of the author.

Alissa Valles: "Mathematician" from *Poetry*, June 2002. Used by permission of the author.

Ramón López Velarde: "Loyal, My Heart," from *Treasury of Mexican Love Poems, Quotations & Proverbs*, edited and translated by Enriqueta Carrington, Hippocrene Books, copyright © 2003 by Enriqueta Carrington. Used by permission of the translator.

John Vieira: "The Lake Swan, the Tom" from *AGNI 22& 56*, Boston University, www.bu.edu/agni/poetry/print/2002/56-vieira.html. Used by permission of the author.

Andrei Voznesensky: "Parabolic Ballad," translated by W.H. Auden, from *Antiworlds and the "Fifth Ace": Poetry by Andrei Voznesensky*, edited by Patricia Blake and Max Hayward, Anchor Books, Doubleday & Company, Inc., copyright © 1963 by Encounter Ltd., copyright © 1965, 1966 by Basic Books, Inc. Reprinted by permission of Basic Books, a member of the Perseus Books Group.

William Wall: "Mathematics: Insects" from *Mathematics and Other Poems*, by William Wall. The Collins Press, Cork, Ireland, copyright © 1997 by William Wall. Used by permission of the author.

Ronald Wallace: "Chaos Theory" from *Ploughshares*, Winter 1995–1996. Used by permission of the author.

Ellen Wehle: "The Song of 10" first appeared in *Epoch and Poetry International 2005*, and was featured on *Poetry Daily*. Used by permission of the author.

John Wheelwright: "Two Tongues in a Tower" from *Collected Poems of John Wheelwright*, copyright © 1971 by Louise Wheelwright Damon. Reprinted by permission of New Directions Publishing Corp.

*William Carlos Williams:* "Aigeltinger" from *Collected Poems 1939–1962, Volume II,* copyright © 1948 by Williams Carlos Williams. Reprinted by permission of New Directions Publishing Corp, and Carcanet Press Limited.

*Clarence* R. *Wylie,* Jr.: "Paradox" from *The Scientific Monthly* 67, July 1948. Used by permission of C. Victor Wylie.

*Yang Mu:* "Geometry: River Goddess" translated by Michelle Yeh and the author, from *The Columbia Anthology of Modern Chinese Literature,* ed. Joseph S. M. Lau and Howard Goldblatt, Columbia University Press, 1995. Used by permission of the author and translators.

# Index

## A

*The Accounting*, 16
Ackerman, Diane, 3
*The Acrobat*, 13
*Addition [1]*, 99
Adherent of convergence, I find solace, 104
*Aigeltinger*, 197
Alberti, Rafael, 73, 127
Alcosser, Sandra, 128
Alegre, Manuel, 4
Alenier, Karren LaLonde, 75
*Alexander's Horned Sphere*, 178
Alighieri, Dante, 76
Along a parabola life like a rocket flies, 63
*alphabet: love*, 14
Amichai, Yehuda, 5
Among the colors, green, we are the sudden, 24
And let us not confine our cares, 189
Andre, Carl, 77
Apostolescu, Stelian, 9
*Argument*, 179
As a boy, I'd still have asked, 111
As he cleaned the board, 179
*Asymptotes*, 115
At a hard wood desk, on a hard, 85
At twenty, I got drunk on Friday nights, 194
Auden, W. H., 64

## B

Baby and you, 120
Back in the iron tower he said, "When I left the tower, 68
*Barbie Says Math Is Hard*, 111
Barry, Amy Quan, 6
Baumel, Judith, 7

The Beaver brought paper, portfolio, pens, 134
Because the zed, 55
Before curves kiss, 133
Being a happy deterministic man, 161
Bellos, David, 131
Bergmann, S. J., xii
Bernoulli, Jakob, 130
Bhaskaracharya, 131
Bierds, Linda, 79
Blaga, Lucian, 9
Bloch, Ariel, 57
Bloch, Chana, 5, 57
Bobo, Ray, 132
*Borderline—A Fractal Poem*, 186
Brooks, David, 10
Browning, Elizabeth Barrett, 11

## C

Calandro, Ann, 133
Calculating with exactitude, 143
*Calculus*, 153
Call it windfall, 7
Carrington, Enriqueta, 123
Carroll, Lewis, 134
*Cat in a Box*, 23
Catullus, Gaius Valerius, 12
Cedering, Siv, 13
*Chaos Theory*, 66
Chapman, Robin, 80
*Child Margaret*, 112
The child Margaret begins to write numbers on a Saturday morning, the first numbers formed under her wishing child fingers, 112
Chipman, Edward E., 81
Christensen, Inger, 14
Ciardi, John, 76
Cibulka, Hanns, 136
Clark, Dan, 82
Cohen, Marion Deutsche, 137, 138
Come, let us hasten to a higher plane, 38
*The Comet*, 103
*The Computation*, 21
Concentrical, the universe and I, 102

Cornish, Mary, 139
Coulton, Jonathan, 141
*Counting* (Douglas Goetsch), 156
*Counting: Gregor Mendel in the Prelacy,* 79
*Counting* (Philip Larkin), 37
Crow, Jennifer, 83
Cunningham, J. V., 15
*The Cyberiad,* 38

D

Daniels, Keith Allen, 143
Davis, Jon, 16
*The Definition of Love,* 42
Dempster, E. P., 145
*The Departure of an Alphabet,* 97
Deppe, Theodore, 146
de Sponde, Jean, 17
DeZur, Kathryn, 85
*Dialectic of the Census Takers,* 75
Dickey, James, 148
Dickinson, Emily, 18, 19
*Distances,* 48
Donne, John, 20, 21
The door is closed, 77
Dove, Rita, 86, 149
Do you have the umbrella?, 73
Drawn by the window-light, two insects, 65
Drost, John L., 150

E

Early, Gerald, 22
Enheduanna, xi
The enthusiasm with which I repeatedly declare you my one, 44
*Entropy,* 26
*An Equation for My Children,* 108
Essinger, Cathryn, 87, 151, 152
Even as the finite encloses an infinite series, 130

F

Father, as old as you when I was four, 122
*Fibonacci,* 7
*Fibonacci Numbers,* 85
*Figures of Thought,* 172

*Five Poems about Zero,* 55
*Flash Cards,* 86
*Floating,* 25
Foncannon, Jet, 180
For months I had been "grinding" Mathematics day and night, 145
For pairs of lips to kiss maybe, 188
For the first twenty years, since yesterday, 21
For want of a nail the shoe was lost, 66
*The Freezing Point of the Universe,* 56
From eye to object no straight line is drawn, 101
From the Romans' *decem* our decibels and decimal system, O tenfold, 196
Frost, Robert, 89

G

Galvin, James, 90
Gardner, Martin, xiii
*Geese,* 109
*Geometry Is the Mind of God,* 90
*Geometry of Moods,* 102
*Geometry* (Rita Dove), 149
*Geometry: River Goddess,* 69
*Geometry* (X. J. Kennedy), 34
*Give Me an Epsilon and I Will Treat It Well,* 132
"Give me a place to stand," Archimedes said, 17
Glaz, Sarah, 91, 153, 155
*God Loves a Curve,* 81
Goetsch, Douglas, 156
Gosset, Thorold, 189
Granville, Jennifer, 23
*Gravity & Levity,* 51
Greer, Mick, 4
Grosholz, Emily, 24, 92, 94
Growney, JoAnne, 25, 158, 160
Grumman, Bob, 95
Guo, Haipeng, 161

H

Hammel, Miriam, 26
Har, Israel, 96
*Harold Lloyd, Student,* 73
Hass, Robert, 163
He drew a circle that shut me out—, 107

The heron resolves itself from the gray lake the water, 51
He stands, a boy with hands that tremble, 83
He's teaching her arithmetic, 82
He was born to wonder about numbers, 185
His lady was lovely, with glamour galore, 165
Hix, H. L., 27
Hoagland, Tony, 29
Hoffman, Roald, 31
Holden, Jonathan, 32, 97
How do I love thee? Let me count the ways, 11
*How to Find the Longest Distance Between Two Points*, 101
How true is the line that keeps threading its way through, 61
Hughes, Langston, 99
Hunter, J. A. H., 165
*The Hunting of the Snark: The Beaver's Lesson*, 134
The hypothesis: I am tense and so are you, 49

## I

I am obliged to plant a grove, 193
Ibn Gabirol, Solomon, 100
I can still see the curve I first traced from the equation, 115
I could live like this, waiting on the roof, 146
I do not know what colour the ships are, 4
I do not understand the poets who tell me, 87
I'd walk close to buildings counting, 156
*I Even Know of a Mathematician*, 150
If a man understands a poem, 117
*If* $dx/dy = (4x^3 + x^2 - 12)/(\sqrt{(2x^2 - 9)})$*, then*, 6
If I ever write about you—, 91
*I Find My Faith in the Flatness of Space*, 104
If n is an even number, 54
I like the generosity of numbers, 139
In classical flower arrangement, 92
I never thought that you'd go beyond, 23
In math I was the whiz kid, keeper, 86
In school I was attracted, 47
In the bare trees old husks make new designs, 197
In your super-logical, 152
I prove a theorem and the house expands, 149
I remember a problem in a math book, 5
I scarce believe my love to be so pure, 20
*Israeli Travel: Otherness is All, Otherness is Love*, 5

I stole forth dimly in the dripping pause, 89
I tally in the ancient way, 9
I tell my students the story of Newton versus Leibniz, 153
It has something to do, 31
It may be esoteric and perverse, 108
It's all I have to bring today—, 18
I used to speak in anagrams during sex no wonder, 56

J

*July 18, 2005*, 50
Just before she flew off like a swan, 29

K

KANDEL, MICHAEL, 39
KATZ, LISA, 109
KENNEDY, X. J., 34
*The Kingly Crown: The Praises of God*, 100
KIRKUP, JAMES, 101
*The Kiss Precise*, 188
*The Kiss Precise (Generalized)*, 189
KLINE, A. S., 12
KRONFELD, CHANA, 5
KUNITZ, STANLEY, 102
KUROSAKA, BOB, 166

L

*A Lady and Her Lingerie*, 165
*A Lady of 80*, 170
A lady of 80 named Gertie, 170
*The Lake Swan, the Tom*, 124
The lamp-light falls on blackened walls, 169
LAPIDUS, JACQUELINE, 35
LARKIN, PHILIP, 37
Last night in the dislabor, 75
*Late Afternoon at the Workshop on Commutative Rings*, 155
*A Lecture on Thomson's Galvanometer*, 169
LEHRER, TOM, 167
LEM, STANISLAW, 38
*Let's Live and Love: To Lesbia*, 12
Let us live, my Lesbia, let us love, 12
*Let Us Now Praise Prime Numbers*, 190
LEVIN, GABRIEL, 96
LEW, JOHN STEPHEN, 103

Lewis, Bernard, 100
Lewis, Israel, 104
*Lilavati,* 131
Lok, Henry, 105
love exists, love exists, 14
The Lover Wishes to Be of Unending, 119
Love = lim_ego→0(1/ego), 168
*Love's Growth,* 20
*Love Story,* 91
*Loyal, My Heart,* 123
Loyal, my heart gains merit in shadows, 123

## M

Maddox, Marjorie, 40
Making love we assume, 32
*Mandelbrot Set,* 141
Margarido, Graça, 4
Markham, Edwin, 107
Marvell, Andrew, 42
Maslanka, Kaz, 168
*Math* (James Dickey), 148
*Math* (Wendy Mnookin), 110
*Mathemaku No. 10,* 95
*A Mathematical Madness,* 145
*Mathematician,* 194
*The Mathematician, Counting,* 151
A mathematician fell madly in love, 181
*The Mathematician in Love,* 181
A mathematician left the convention, 160
A mathematician was obsessed with things prime, 150
*Mathematics* (Hanns Cibulka), 136
*Mathematics: Insects,* 65
*Mathematics* (Jennifer Crow), 83
Mathews, Harry, 44, 46
*Math Is Beautiful and So Are You,* 54
*Mathrimony,* 143
Maxwell, James Clerk, 169
Maxwell said: There is no more powerful way, 191
McClellan, John Ward, 170
*The Meaning of Zero: A Love Poem,* 60
Meinke, Peter, 47, 48
A mere eyelid's distance between you and me, 60

*Midpoint: The Play of Memory*, 122
MIKHAIL, DUNYA, 49
MILLS, WILMER, 108
MISHOL, AGI, 109
MNOOKIN, WENDY, 110
*Moon Compasses*, 89
MORI, KYOKO, 111
MUNROE, RANDALL, 171
My best dream is floating, 25
My companions since childhood, these numbers, 79
My dad gave me one dollar bill, 113
*My Dance is Mathematics*, 158
*My Dog Practices Geometry*, 87
My father tested as a genius, 97
My love is of a birth as rare, 42
My math teacher Epstein, 109
My mother was burning in a closet, 163
My new father doesn't like it either, 110
*My Number*, 128
My number is small. A hundred pounds of water, 128

N

The nature of a faithful content ideal, 155
Near from infinity I came, 103
NEMEROV, HOWARD, 172
NERUDA, PABLO, 173
*The New Poetry Handbook*, 117
NIED, SUSANNA, 14
NIKAIDO, DEANNA, 50
*Nonlinear Function*, 80
A nonlinear life doesn't turn out, 80
nothing's been quite the same with me, 35
Not truth, nor certainty. These I forswore, 198
*Number Man*, 185
The numbers, 136
*Numbers*, 139
Numbers scraping their heels all night in the attic above them, 16
Numbers who can't ever hear me, 148

O

O'BRIEN, KATHARINE, 176, 177
October she phoned to say, 58

*Ode to Numbers*, 173
Oh, the thirst to know, 173
O Light Eternal fixed in Itself alone, 76
On a connected surface, 137
One cannot help loving a mathematician, 151
*On the Sadness*, 77
*On Your Imminent Departure: Considering the Relative Importance of Various Motions*, 53
*Orders of Magnitude*, 27
OSERS, EWALD, 136
*Outwitted*, 107

P

*Parabolic Ballad*, 63
*The Paradiso: Canto XXXIII*, 76
*Paradox*, 198
PARRY, BILL, 178, 179
Pathological monsters! cried the terrified mathematician, 141
PEDEN, MARGARET SAYERS, 175
poetry ⟌existence, 95
A point is a point, a line is a line, 176
A point is that which has no part, 90
*Polar Heart*, 4
*Practical Application*, 82
*The Proof*, 146
*Proportions of the Heart*, 92
*Puzzle*, 193

Q

QUENEAU, RAYMOND, 180

R

RAMKE, BIN, 51
RANKINE, WILLIAM J. MACQUORN, 181
*The Ratio of Green*, 24
ROBERTS, LEN, 183
ROGERS, PATTIANN, 53

S

*Sacrifice and Bliss*, 168
*Safety in Numbers*, 44

SAKELLARIOU, BECKY DENNISON, 54
SALVAGGIO, ERYK, 55
*San Antonio, January, 1993*, 160
SANDBURG, CARL, 112, 185
*Scared and the Intermediate Value Theorem*, 137
SEATON, MAUREEN, 56
The sea was the first to propose it, 178
*Self-Improvement*, 29
7 × 7 + love =, 99
*Several Hypotheses and a Proposition*, 35
*Sex and Mathematics*, 32
SEYKOTA, ED, 186
*Shakespeare on Love*, 46
Shall I compare thee to a summer's day, 46
*The Shape of Desire*, 94
The shape of her soul is a square, 114
*She Considers the Dimensions of Her Soul*, 114
She said that entropy, 26
She who arrives across the water, 69
Show me a sign, 50
SILVERSTEIN, SHEL, 113
*Sine Qua Non*, 116
SJÖBERG, ÅKE W., xii
SLAVITT, DAVID R., 17
*Smart*, 113
SMITH, YOUNG, 114
SOCOLOW, ELIZABETH ANNE, 115
SODDY, FREDERICK, 188
SOLOMON, KING, 57
Some distances cannot be crossed; like, 48
*The Song of 10*, 196
*The Song of Songs*, 57
*Sonnet of Love XIII*, 17
SPALDING, HELEN, 190
A Square in verſe of a hundred monaſillables only, 105
$\sqrt{\heartsuit}$=?, 171
STALLINGS, A. E., 116
"Stand," she said, in the night air, "stand still as a star," 22
*Statistics*, 15
STEAD, C. K., 58
STRAND, MARK, 74, 117
STRICKLAND, STEPHANIE, 119, 120, 191

*Striving All My Life*, 191
STUEBEN, MICHAEL, 192
*Sundry Christian Passions: Square Poem in Honor of Elizabeth I*, 105
*Swigns*, 180

T

*Tales from A Sonnetarium: III, Kismet*, 3
*Tally*, 9
*The Theory of Absence*, 49
There is no bravery in flight, 13
*There's a Delta for Every Epsilon*, 167
There, stuck in that class, 40
*There Was a Young Maiden*, 166
There was a young maiden named Lizt, 166
They both collected for the March of Dimes, 15
They called you der Noether, as if mathematics, 158
They say who play at blindman's buff, 34
Thinking in terms of one, 37
Thou art One, the beginning of all computation, the base of all construction, 100
*3.141592 . . .*, 47
Threescore are the queens, 57
TIPTON, CAROLYN L., 127
*To Divine Proportion*, 127
To lay the logarithmic spiral on, 172
To you, amazing discipline, 127
Tracing an airplane's pale trajectory, 94
*Treatise on Infinite Series*, 130
*True North 1*, 119
*Two Tongues in a Tower*, 68

U

*Undefined Terms*, 176
*Unfinished Calculations*, 61
UNKNOWN AUTHOR, 193
UPDIKE, JOHN, 122
*Useless*, 171
UYEMATSU, AMY, 60, 61

V

*Valentine* (Katharine O'Brien), 177
*Valentine* (Michael Stueben), 192

Valles, Alissa, 194
Velarde, Ramón López, 123
*Venn Diagrams,* 40
Vieira, John, 124
Voznesensky, Andrei, 63

W

Walker, Brenda, 9
Walker, Helen M., 130
*Walking Westward,* 58
Wallace, Ronald, 66
Wall, William, 65
Wehle, Ellen, 196
Weiner, Joyce S., 180
*We Sat, So Patient,* 183
We sat, so patient, in that third, 183
*We shall find the Cube of the Rainbow,* 19
What drove me into math, 138
Wheelwright, John, 68
*When a P-Man Loves an NP-Woman,* 161
When nothing remains, 96
when One made love to Zero, 180
When Thales learned to measure in his head, 27
*Where the Kissing Never Stops,* 133
Which is more important, the motion of the wind, 53
Whilst making love a necklace broke, 131
*"...White of Forgetfulness, White of Safety",* 163
*Who Counts, Counts,* 120
*Why Does Disorder Increase in the Same Direction of Time as That in Which the Universe Expands?,* 31
Williams, William Carlos, 197
Winslow, Elizabeth, 49
*With Linnet and Rosalind on the Evening of the Equinox (or an Exposition on the Big Bang Theory),* 22
Wittgenstein was wrong: when lovers kiss, 3
Wylie, Clarence R., Jr., 198

X

x = Ø Found alone and wanting, 132

Y

Yang Mu, 69
Yeh, Michelle, 70

*Yes,* 10
*You Are Right,* 152
you are standing at the ocean, 6
You are the fairest of your sex, 192
You disintegrate my differential, 177
You know how, 10
Your absence, father, is nothing. It is naught—, 116

Z

*The Zero,* 96
Z-naught, a point is squared and when, 186